Numerical Modelling of Bulk Superconductor Magnetisation

Numerical Modelling of Bulk Superconductor Magnetisation

Mark Ainslie
Department of Engineering, University of Cambridge, UK

Hiroyuki Fujishiro
Department of Physical Science and Materials Engineering, Iwate University, Japan

IOP Publishing, Bristol, UK

Supplementary COMSOL files to accompany this book are available at http://iopscience.iop.org/book/978-0-7503-1332-2.

ISBN 978-0-7503-1332-2 (ebook)
ISBN 978-0-7503-1333-9 (print)
ISBN 978-0-7503-1957-7 (myPrint)
ISBN 978-0-7503-1334-6 (mobi)

DOI 10.1088/978-0-7503-1332-2

Version: 20191101

IOP ebooks

British Library Cataloguing-in-Publication Data: A catalogue record for this book is available from the British Library.

Published by IOP Publishing, wholly owned by The Institute of Physics, London

IOP Publishing, Temple Circus, Temple Way, Bristol, BS1 6HG, UK

US Office: IOP Publishing, Inc., 190 North Independence Mall West, Suite 601, Philadelphia, PA 19106, USA

Contents

Preface vii

Acknowledgements ix

Author biographies x

1 Fundamentals of bulk superconducting materials 1-1

1.1 Bulk superconductors 1-1
1.2 Magnetic properties of bulk superconductors 1-3
1.2.1 Superconducting material classifications 1-3
1.2.2 Flux pinning and trapping 1-4
1.2.3 Flux creep 1-5
1.3 Fabrication processes 1-6
1.3.1 Bulk (RE)BCO superconductors 1-6
1.3.2 Bulk MgB_2 superconductors 1-8
1.3.3 Bulk iron–pnictide superconductors 1-9
1.4 Magnetisation of bulk superconductors 1-9
1.4.1 Pulsed field magnetisation 1-11
1.5 Bulk superconductor applications 1-12
1.5.1 Flux pinning applications 1-13
1.5.2 Flux trapping applications 1-14
1.5.3 Flux shielding applications 1-17
1.5.4 Magnetic lens 1-17
1.5.5 Conductor alternative 1-18
References 1-18

2 Numerical modelling of bulk superconducting materials 2-1

2.1 Modelling of bulk superconductors 2-1
2.1.1 Analytical techniques 2-2
2.1.2 Numerical techniques 2-4
2.2 Finite element method (FEM) 2-4
2.2.1 Modelling bulk superconductors using FEM 2-5
References 2-13

3 Modelling magnetisation of bulk superconductors 3-1

3.1 Magnetisation of bulk superconductors 3-1

3.1.1 Zero-field-cooled (ZFC) and field-cooled (FC) magnetisation 3-1
3.1.2 Pulsed field magnetisation (PFM) 3-17
References 3-39

4 Demagnetisation and novel, hybrid superconductor structures 4-1

4.1 Demagnetisation effects and AC losses 4-1
4.2 Novel and hybrid bulk superconductor structures 4-10
4.2.1 Composite structures with improved thermal conductivity 4-10
4.2.2 Hybrid ferromagnet-superconductor structures 4-12
4.2.3 Hollow bulk cylinders and tubes for shielding 4-20
4.2.4 Hybrid trapped field magnet lens 4-24
References 4-31

Appendix A A-1

Preface

At cryogenic temperatures, bulk superconductors can be used as so-called trapped field magnets (TFMs) that can trap magnetic fields an order of magnitude higher than the maximum magnetic field available from conventional permanent magnets, which is limited practically to less than around 2 T, less than the saturation magnetisation of most practical ferromagnetic materials. This makes them highly attractive for a number of engineering applications that rely on high magnetic fields, including rotating electric machines, magnetic resonance imaging (MRI), nuclear magnetic resonance (NMR), magnetic separation, magnetic drug delivery systems (MDDS) and Lorentz force velocimetry.

The techniques used to fabricate bulk superconducting materials have continued to improve and are now at a stage where high performance materials, capable of trapping fields in excess of 17 Tesla, can be grown reliably on a reasonably large scale, and a number of commercial suppliers exist around the world capable of producing in excess of 1000 bulk superconductors per year. There are also a number of conceptual designs and successful prototypes of bulk superconductor-based devices that have been developed by researchers and industry, bringing together an environment where the technology can be exploited commercially in a variety of fields in the near future.

Numerical modelling is a particularly important and cost-effective method to guide both superconducting material processing and practical device design, and has been used successfully to date to interpret experimental results and the physical behaviour and properties of the bulk superconductor during its magnetisation process, to propose and analyse new magnetisation techniques, and to design and predict the performance of bulk superconductor-based devices.

In this book, we provide readers with numerical analysis techniques to model the magnetisation of bulk superconductors based on the finite element method (FEM). The initial focus is on the electromagnetic properties of such materials, which is sufficient for some cases; however, techniques to model both the electromagnetic and thermal properties are also introduced, such as numerical simulation of pulsed field magnetisation (PFM), where the temperature of the bulk superconductor can rise significantly due to the rapid, dynamic movement of magnetic flux within the material during PFM.

The text is suitable for undergraduate and postgraduate students, as well as engineers and scientists with an interest in understanding and modelling the magnetisation properties of bulk superconducting materials and/or exploiting such materials as super-strength, high field magnets in engineering applications.

Chapter 1 of the book provides the reader with a general overview of practical bulk superconducting materials, including their magnetic properties, their various compositions and fabrication processes, how they are magnetised, and some of the engineering applications in which their remarkable properties can be exploited, which include flux pinning, flux trapping and flux shielding applications.

Chapter 2 introduces some of the analytical and numerical modelling techniques that can be used to analyse the magnetisation properties of bulk superconductors and provides an overview of the general process used to model such materials using FEM. An overview of modelling the electrical and thermal properties of bulk superconducting materials, including how to couple these together, is provided. This is particularly important as the superconductor's electrical properties vary with temperature and under certain magnetisation processes and conditions, the bulk can experience significant changes in temperature.

In order to utilise a bulk superconductor as a strong TFM, they must be magnetised, which essentially involves the application and removal of a large magnetic field that induces a circulating supercurrent in the material that flows without resistance. Chapter 3 details how the modelling frameworks described in chapter 2 can simulate the different magnetisation techniques for bulk superconductors: zero-field-cooled (ZFC) and field-cooled (FC) magnetisation, which are typically used to determine the maximum trapped field capability of a bulk superconductor, and pulsed field magnetisation (PFM), which shows great promise as a compact, mobile and relatively inexpensive magnetisation technique for magnetising bulk superconductors in practical applications. The influence of the electrical and thermal properties on the magnetisation of bulk high-temperature superconductors is also discussed, as well as other bulk superconducting materials, such as MgB_2 and iron-pnictide materials.

Chapter 4 discusses the modelling of some further practical considerations for bulk superconductors, such as the effect of externally applied magnetic fields in different configurations on the attenuation of the trapped field (demagnetisation), which is crucial in applications like rotating machines where the superconductor can experience an external, time-varying magnetic field. A number of novel and hybrid bulk superconductor structures are also discussed, including improved thermal conductivity structures and ferromagnet-superconductor structures, which have been designed to overcome some of the issues related to bulk superconductors and their magnetisation and enhance the intrinsic properties of bulk superconductors acting as TFMs. Finally, the use of hollow bulk cylinders and tubes for shielding applications is discussed, as well as a new concept of a hybrid trapped field magnet lens.

Throughout the book, a number of example models, implemented in the FEM software package COMSOL Multiphysics®, are provided to the reader so that they may utilise these tools to analyse specific cases of interest and to develop their own models. We hope that this book provides the reader with some useful insights into the numerical modelling of bulk superconductor magnetisation towards their practical use in engineering applications.

Mark Ainslie and Hiroyuki Fujishiro
August 2019

Acknowledgements

Dr Mark Ainslie would like to acknowledge financial support from a Royal Academy of Engineering Research Fellowship and an Engineering and Physical Sciences Research Council (EPSRC) Early Career Fellowship EP/P020313/1.

Author biographies

Mark Ainslie

Dr Mark Ainslie received the BE (Electrical and Electronic) and BA (Japanese) degree from the University of Adelaide, Australia, in 2004, the MEng degree from the University of Tokyo, Japan, in 2008, and the PhD degree from the University of Cambridge, UK, in 2012. In 2011, he was awarded the European Society for Applied Superconductivity (ESAS) Young Researcher's Award in Large Scale Applications for his PhD work on transport AC loss in high-temperature superconducting (HTS) coils. He is also co-holder of a Guinness World Record for the 'Strongest magnetic field trapped in a superconductor' (https://www.guinnessworldrecords.com/world-records/439929-strongest-magnetic-field-trapped-in-a-superconductor/).

From 2012 to 2017, he was a Royal Academy of Engineering Research Fellow in the Bulk Superconductivity Group in the Department of Engineering, University of Cambridge, where he investigated the engineering interactions of conventional, magnetic and superconducting materials for electrical applications. This work focussed on the use of HTS materials in bulk and wire form to increase the electrical and magnetic loadings of an axial gap, trapped flux-type superconducting electric machine. In 2017, he became an Engineering and Physical Sciences Research Council (EPSRC) Early Career Fellow, also in the Bulk Superconductivity Group, investigating the use of bulk HTS materials in portable, high field magnet systems.

His research interests cover a broad range of topics in applied superconductivity, including superconducting electric machine design, bulk superconductor magnetisation, numerical modelling, and interactions between conventional and superconducting materials.

Hiroyuki Fujishiro

Professor Hiroyuki Fujishiro received the BE (Electrical and Electronic) degree from Tohoku University, Sendai, Japan, in 1980, the MEng degree from Tohoku University, Sendai, Japan, in 1982, and the PhD degree from Tohoku University, Sendai, Japan, in 1985.

From 1985 to 1989, he was a researcher at the Semiconductor Research Institute, Sendai, Japan, where he investigated the crystal growth of III-V compound semiconductors. From 1989 to 1991, he was Assistant Professor at Kyushu Institute of Technology, Fukuoka, Japan, where he mainly investigated the crystal growth of the semiconductor thin films. From 1991 to 1993, he was Assistant Professor at Iwate University, Morioka, Japan, where he started the study of thermal properties of HTS materials. From 1993 to 2006, he was Associate Professor. In 2003, he was a visiting researcher at the Institute of Physics of ASCR, Prague, Czech Republic. In 2005, he was awarded the Best Presentation Award from the Cryogenics and Superconductivity Society of Japan for his work on 'Thermal properties of REBCO bulk superconductors (RE=Dy, Gd, Ho, Nd)'. In 2006, he became Professor at Iwate University. In 2019, he was promoted to Vice President/Executive Director for Research, Revitalisation and Regional Development of Iwate University.

His research interests cover a broad range of topics in applied superconductivity, including experiments on bulk superconductor magnetisation (mainly pulsed field magnetisation and field-cooled magnetisation) and the numerical simulation of electromagnetic, thermal and mechanical behaviours during these magnetising processes.

IOP Publishing

Numerical Modelling of Bulk Superconductor Magnetisation

Mark Ainslie and Hiroyuki Fujishiro

Chapter 1

Fundamentals of bulk superconducting materials

Chapter 1 of this book provides the reader with a general overview of practical bulk superconducting materials, including their magnetic properties, their various compositions and fabrication processes, how they are magnetised, and some of the engineering applications in which their remarkable properties can be exploited, which include flux pinning, flux trapping and flux shielding applications.

1.1 Bulk superconductors

Bulk superconductors, acting as trapped field magnets (TFMs), can be used, when cooled to cryogenic temperatures, as super-strength, stable permanent magnets generating fields of several tesla. When compared to the 1.5–2 tesla limit for conventional permanent magnets, such as neodymium (Nd–Fe–B) magnets, this makes them attractive for a number of engineering applications that rely on high magnetic fields, including compact and energy-efficient electrical machines with unprecedented power densities and compact and portable magnetic resonance imaging (MRI) and nuclear magnetic resonance (NMR) systems. It is also possible for scientists to use high magnetic fields to exploit the magnetism of a material for controlling chemical and physical processes, which is attractive for magnetic separation and magnetic drug delivery systems (MDDS), for example. With the continued development of conventional (wire-wound) superconducting magnets and the achievement of higher magnetic fields, even the chemical and physical processes associated with diamagnetic materials, which make up many of the materials found on earth, are significantly influenced. This has led to observations of the Moses effect (where the free surface of water is deformed by a magnetic field of several tesla), magnetic levitation of diamagnetic materials, and magnetic orientation of organic polymers and gels and carbon nanotubes, which is particularly attractive for improving crystal growth of organic semiconductors and other materials.

doi:10.1088/978-0-7503-1332-2ch1

However, one significantly challenging problem currently faced is achieving a simple, reliable and portable charging technique to magnetise such superconductors, and this is crucial to harnessing their remarkable properties to develop competitive and compact designs for high field, trapped flux-type superconducting applications, such as electric machines. The magnetisation process of a bulk superconductor essentially involves the application and removal of a large magnetic field that induces a circulating supercurrent in the material that flows without resistance. The University of Cambridge's Bulk Superconductivity Group research team recently broke the world record for the trapped magnetic field in a single-grain, bulk high-temperature superconducting (HTS) material with 17.6 T at 26 K [1], leapfrogging the previous record of 17.24 T at 29 K [2] that stood for over a decade, highlighting the significant potential of such materials. In addition, the results from this new world record showed that these materials are capable of trapped fields greater than 10 T at around 50 K, a temperature that is readily achievable with commercial, single-stage cryocoolers.

A comparison of a conventional permanent magnet (PM), a bulk superconductor and an electromagnet, which could be conventional (copper-wound) or superconducting, is shown in figure 1.1, highlighting the main differences in the magnetic field available in terms of shape and magnitude, and the mechanism by which each magnet generates its magnetic field. The chief advantage of a bulk superconductor as a TFM is that the available field can be up to an order of magnitude higher than conventional PMs and no power supply and direct connection is necessary to supply the current producing the magnetic field, as in electromagnets. Such TFMs can provide a compact, stationary and strong magnetic field with a low stray field compared with wire-wound magnets.

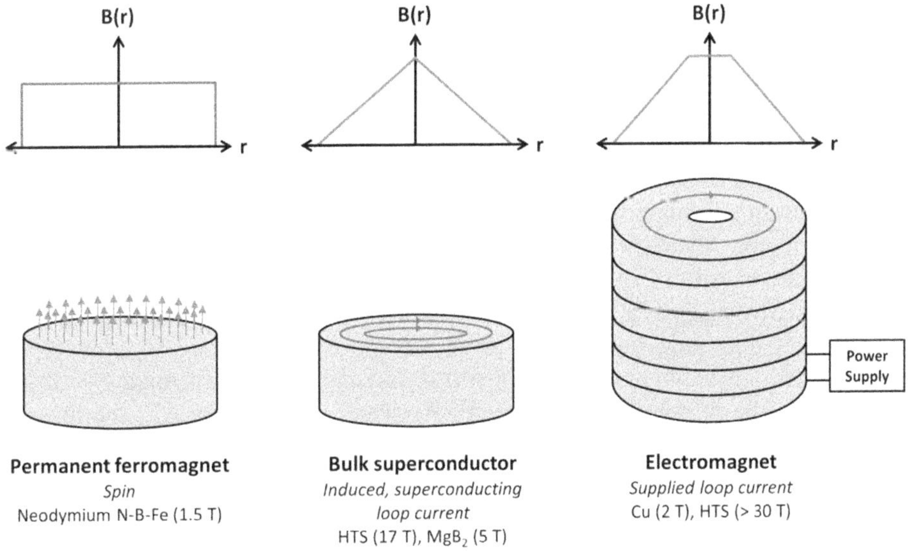

Figure 1.1. A comparison of the magnetic field available from a conventional permanent magnet (PM), a bulk superconductor and an electromagnet, and the mechanism by which each magnet generates its magnetic field.

This chapter provides the reader with a general overview of practical bulk superconducting materials, including their magnetic properties, their various compositions and fabrication processes, how they are magnetised, and some of the practical applications in which they can be exploited.

1.2 Magnetic properties of bulk superconductors

1.2.1 Superconducting material classifications

1.2.1.1 Low- and high-temperature superconducting materials

Superconducting materials are generally classified in two ways. Firstly, by temperature, leading to definitions of low-temperature superconducting (LTS) and high-temperature superconducting (HTS) materials. LTS typically refers to the Nb-based alloys (commonly Nb47wt%Ti) and A15 type superconductors (Nb_3Sn, Nb_3Al) that are still used in many commercial applications in wire form, such as MRI, high field and accelerator magnets. The LTS materials are typically also known as conventional superconductors, whose properties can be fully explained by BCS theory [3] and other related theories, and have a critical temperature, T_c, below 30 K. HTS materials refer to superconducting materials with T_c above 30 K that cannot be explained by conventional superconductivity theories and are commonly thought of as superconducting materials with a T_c greater than 77 K, the boiling point of liquid nitrogen. The newly discovered magnesium diboride (MgB_2) superconducting system has a T_c around 39 K and is typically utilised at an operating temperature of 20 K or lower, but cannot be explained by conventional theories; hence, its categorisation based on these definitions is difficult.

1.2.1.2 Type I and II superconductivity

Another classification is based on the magnetic response of the superconductor to an externally applied magnetic field, leading to definitions of type I and type II superconductors. Type I superconductors have only two states, while there are three in a type II superconductor. The characteristic feature of a type I superconductor is the Meissner effect (perfect diamagnetism) below a critical field, H_c, which implies the existence of a surface energy boundary between any normal and superconducting regions in the material. When the external field exceeds H_c, the type I superconductor reverts back to the normal state, where the field completely penetrates the material. Compared with a type I superconductor, there is a 'mixed' state (or mixed phase) in a type II superconductor, where there exists a lower and upper critical magnetic field: H_{c_1} and H_{c_2}, respectively. Below H_{c_1}, the type II superconductor behaves as a type I superconductor and any magnetic field is expelled from the interior of the material. Above H_{c_1}, but below H_{c_2}, the magnetic field penetrates partially into the material in the form of vortices (or magnetic flux lines or quantum 'flux tubes'), within which the material is in the normal state, and outside remains in the superconducting state. In this mixed state, the material is threaded by vortices with flux parallel to and in the same direction as the applied field, H_a. The resistivity of a superconductor may be vanishing in the mixed state, provided the vortices are pinned or trapped (see section 1.2.2). Each quantum of flux, $\Phi_0 = h/2e$, is surrounded by a circulating 'screening' supercurrent and the interaction between each vortex current and magnetic fields from other vortex currents results in a repulsive force that results in the cores arranging themselves into a periodic array called the Abrikosov vortex lattice. With increasing H_a, the normal state would eventually occupy the entire material and above the upper critical field, H_{c_2}, the material becomes completely normal.

The firstly discovered superconductors—metallic elements and alloys, and type I materials—had extremely low T_c and H_c values that severely limited their use in any practical applications. Type II superconductors can carry much larger amounts of current in higher magnetic fields in comparison with type I superconductors because H_{c_2} can be hundreds of times larger than H_c. For example, the strongest type I superconductor (pure lead) has a critical field of around 80 mT (800 Gauss), whereas the well-known, type II superconductor Y–Ba–Cu–O (YBCO) has an H_{c_1} around 20 mT (200 Gauss) and an H_{c_2} as high as 100 T.

1.2.1.3 Irreversibility field

Another defined field for type II superconductors—and one of particular importance for practical applications—is the irreversibility field, H_{irr}. For applied magnetic fields above H_{irr}, the pinning (see section 1.2.2) in the superconducting material becomes ineffective because the Lorentz force—given, as a force per unit volume, by $\boldsymbol{J} \times \boldsymbol{B}$—is greater than pinning force and the vortices are free to move, which also creates additional dissipation. The magnetic behaviour of the superconductor then becomes reversible and is still superconducting, but unable to support a useful current density. H_{irr} is considered the practical limit of type II superconductors and is an order of magnitude lower than H_{c_2} [4], and represents, in the case of bulk superconductors, an upper limit to the field trapping capability of the material. H_{irr} is temperature-dependent, increasing with decreasing temperature, and for YBCO, for example, H_{irr} can be as high as 5–6 T at 77 K [5]. The RE–Ba–Cu–O (where RE = rare-earth elements, such as Gd, Sm and Nd, or Y, and hereafter referred to as (RE)BCO) family of superconducting materials exhibits the highest H_{irr} over a wide temperature range and therefore most promising to carry large currents in high magnetic fields. The term 'bulk HTS' will also refer to these materials in the rest of this book.

1.2.2 Flux pinning and trapping

The ability of a bulk superconductor to 'trap' large magnetic fields is determined by the material's ability to support a large, induced superconducting loop current (see figure 1.1). This current is determined by the magnitude and homogeneity of the critical current density, J_c, and the length scale over which it flows. J_c is the most important characteristic of a superconductor for any practical superconducting device and gives the maximum current density the material can maintain without resistance. In type II materials, or the so-called 'hard' superconductors, J_c exhibits a strong dependence on temperature, as well as local magnetic field, leading to an in-field, temperature-dependent, $J_c(\boldsymbol{B}, T)$, that may also be anisotropic. This behaviour is directly related to the ability of flux pinning sites within the material to effectively pin magnetic flux lines (those described in section 1.2.1.2), which act like a frictional force to prevent the motion of flux lines. The total pinning force that all pinning sites within a unit volume can exert on the flux lines is the pinning force density, F_p, and at J_c, where the Lorentz force is balanced by the pinning force density, $J_c = F_p/B$ [6]. The flux pinning is provided by non-superconducting phases, inhomogeneities and defects within the material, such as dislocations, voids, and grain boundaries. The engineering of such pinning sites into the material's microstructure and increasing their flux pinning strength is an important part of the fabrication process and some techniques for this are discussed in section 1.3.

For a type II superconductor, without the contribution of flux pinning sites, the material will enter the mixed state for an external field above H_{c_1} and the magnetic flux lines are free to move within the material and will arrange themselves into the periodic array described in section 1.2.1.2. This uniform distribution of magnetic field corresponds to a zero magnetic field gradient and no current would flow ($J = 0$); hence, the J_c is effectively zero. If this external field is then removed, these unpinned flux lines are free to leave the material; no magnetic flux remains in the material and hence no remnant or 'trapped' magnetic field. On the other hand, if pinning sites exist within the material, flux lines could be pinned, and a certain gradient of magnetic field would be established, leading to a current $J \neq 0$ and a non-zero J_c. When the external field is removed, flux lines are pinned within the material and an associated superconducting loop current flows in the bulk superconductor. This results in an irreversible hysteresis loop and the so-called 'trapped field'.

1.2.3 Flux creep

The phenomenon of flux creep is observed in type II superconductors, which, in a bulk superconductor, results in an approximately logarithmic decay of the trapped magnetic field with time after magnetisation. At finite temperatures, pinned flux lines can be thermally activated, causing them to hop out of one pinning site to another, driven by the Lorentz force. In [7], Anderson and Kim proposed their flux creep theory, which is summarised in the following equation:

$$J(T, t) = J_{c0}\left[1 - \left(\frac{kT}{U_0}\right)\ln\left(\frac{t}{t_0}\right)\right] \tag{1.1}$$

This model assumes a linear dependence of the effective energy barrier on the current density ($U = U_0(1 - J/J_{c0})$, where U_0 is the barrier height in the absence of a driving force and J_{c0} is the critical current density, in the absence of thermal activation, required to tilt the barrier to zero in this approximation), which gives rise to an exponential I–V curve [8]. U_0 can be determined by measuring the normalised relaxation rate of the magnetisation ($S = \frac{1}{M}\frac{dM}{d(\ln(t))}$). t_0 is the 'effective' hopping attempt time. Equation (1.1) predicts the logarithmic time dependence commonly observed in experiments and the current decay in equation (1.1) is commonly measured through the magnetisation decay [8]. The logarithmic model of Zeldov *et al* [9, 10] assumes $U = U_0 \ln(J_{c0}/J)$, resulting in a power-law I–V relationship, which is one of the core assumptions for bulk superconductor modelling, as detailed in section 2.2.1.3.2. Interested readers are referred to [8] for more information on magnetic relaxation.

1.3 Fabrication processes

The trapped field capability of a bulk superconductor is determined mainly by its ability to maintain a high (and homogeneous) J_c in the presence of large magnetic fields, enabling a large circulating supercurrent to flow over a relatively large area within well-aligned grains of the material. The J_c of the material, as described earlier, is determined by its ability to pin magnetic flux lines under the influence of the Lorentz force and is usually achieved by inhomogeneities in the microstructure that can be engineered through the material's fabrication process.

1.3.1 Bulk (RE)BCO superconductors

In the HTS material family, there are a number of potential candidates: (RE)BCO, Bi–Sr–Ca–Cu–O and Tl–Ba–Ca–Cu–O. H_{irr} for the Bi–Sr–Ca–Cu–O family is quite low (on the order of 100s of mT) and its in-field J_c is comparatively poor, and although Tl–Ba–Ca–Cu–O does exhibit a relatively high H_{irr} at 77 K (around 4 T), thallium is toxic and the material is difficult to process, leading to a relatively low J_c. Hence, much research has focussed on the (RE)BCO family, where RE = typically Y, Gd, Eu, Sm and Nd.

Early attempts at fabricating bulk HTS materials, focussed predominately on YBCO, used simple sintering techniques based on common processing routes for fabricating ceramic materials. Unfortunately these resulted in disappointingly low J_c values associated with granularity and weak-links from high-angle grain boundaries. Melt-processing techniques, based on peritectic solidification, were shown to be effective in avoiding such grain boundaries through better grain alignment, resulting in significantly higher J_c values and enhanced trapped field capability.

A number of such melt-processing techniques have now been developed, which are based on the following peritectic reaction:

$$\begin{array}{cccc} 2(\mathrm{RE})\mathrm{Ba_2Cu_3O_{7-\delta}} \leftarrow & (\mathrm{RE})_2\mathrm{BaCuO_5} + & \mathrm{Ba_3Cu_5O_8} + & x\mathrm{O_2} \\ (123) & (211/422) & \text{Liquid} & \text{Gas} \end{array}$$

The temperature, T_p, at which this peritectic reaction occurs varies with material, e.g., $T_p(\mathrm{YBCO}) \approx 1015$ °C and $T_p(\mathrm{SmBCO}) \approx 1070$ °C, and the so-called top-seeded melt growth (TSMG) is currently the most commonly used fabrication process for bulk HTS. In TSMG, a structurally compatible (same lattice structure) seed with a higher melting point than the bulk pellet (e.g., Nd123, Sm123) is used to 'seed' (or initialise) the large grain growth and control its orientation. This seed promotes heterogeneous grain nucleation, yielding a highly textured structure with its *c*-axis oriented perpendicular to the disc surface, with four facet lines extending from the seed to the edge, reflecting a single grain with a controlled crystal orientation [11].

The overall fabrication process by TSMG can be summarised in four steps: mixing, pressing/seeding, melt-processing and oxygenation. Firstly, precursor powders of a desired composition are mixed together using a mortar and pestle or other mixing device. A couple of hours is a typical mixing time for 200 g of powder. In the case of bulk YBCO, a recommended composition might be 70 wt%Y-123 + 30 wt% Y-211 + 0.1 wt%Pt, where Y-123 is the superconducting phase and Y-211 is the non-superconducting phase. When grown, this then results in a bulk superconducting Y-123 matrix within which discrete non-superconducting Y-211 particles (typically around a few μm) are distributed. The J_c of the bulk is very sensitive to its microstructure and a homogeneous distribution of these fine Y-211 particles correlates with increased flux pinning and higher J_c values. The addition of Pt (or CeO_2) helps inhibit the coarsening of the Y-211 particles to control their growth to yield a larger number of pinning sites. The mixed precursor is then weighed and poured into a die of the desired dimensions, before it is pressed into a pellet. Typical pressures applied are 2–5 tonnes for 20–40 mm diameter pellets. A seed is then placed at the centre of the top surface of the pellet, which is then put into a furnace for melt-processing. In the furnace, the pellet is subjected to a particular temperature profile: firstly the sample is heated above T_p (e.g., 1040° for YBCO), then cooled down to a temperature just above T_p, then very slowly cooled (typically 0.2–0.8 °C h^{-1}) below T_p through the peritectic growth window, which allows the growth of the Y-123 matrix with distributed Y-211 particles. Finally, the samples are oxygenated (annealed in an oxygen atmosphere) to adjust the oxygen stoichiometry of the sample such that $\delta = 0.02–0.07$ and the material undergoes a transformation from a tetrahedral (non-superconducting) crystal structure to orthorhombic (superconducting). The sample can then be machined into the desired shape, polished and used as a TFM.

Refinement of processing techniques has resulted in bulk HTS materials that can exhibit J_c values of 5×10^4 A cm^{-2} at 1 T and 77 K, resulting typically in trapped fields of up to 1–1.5 T for YBCO and greater than 2 T for (RE)BCO at this technologically important temperature, with 3 T in a 65 mm diameter sample the highest trapped field at 77 K so far [12]. Indeed, as described in section 1.1, record magnetic fields exceeding 17 T have been trapped in such materials at temperatures lower than 30 K [1, 2]. It should be noted that these values quoted are with respect to standard, non-irradiated YBCO and (RE)BCO samples, but neutron-irradiated YBCO samples have exhibited higher J_c values [13, 14] and trapped fields over 2 T at 77 K in samples as small as 20 mm in diameter [15, 16].

There were also attempts several decades ago using bulk LTS materials, e.g., in the 1970s. For example, 2.24 T was trapped at 4.2 K in a hollow Nb_3Sn superconducting cylinder, constructed by helically wrapping multiple layers of superconducting 'ribbon' around a mandrel [17]. However, this technique and material has not been applied to practical applications because of problems with thermal instability due to low specific heat, frequent flux jumps and low mechanical strength.

1.3.2 Bulk MgB_2 superconductors

Another superconductor that shows promise in bulk form as a TFM is MgB_2. MgB_2 was first reported in [18] and although the transition temperature for this material is low (39 K), which requires a lower operating temperature (20 K or less) and more complex cryogenic system than that for (RE)BCO bulks, the material is lighter weight and has a more homogeneous J_c distribution. While bulk (RE)BCO materials can have high J_c values that allow record trapped fields, they can suffer from an inhomogeneous J_c distribution within the sample volume and it is difficult to fabricate large, high-quality single-grain samples over 50–60 mm in diameter. The relative ease of fabrication of MgB_2 materials, as well as their long coherence length [19], lower anisotropy and strongly linked supercurrent flow in untextured polycrystalline samples [20, 21], has enabled a number of different processing techniques to be developed, including *in situ* and *ex situ* methods. The formation of MgB_2 is mainly from mixed powders of Mg and B; in a typical *in situ* reaction process, Mg grains melt and diffuse into B grains, but this typically results in low density bulk MgB_2 samples. A hot isostatic pressing (HIP) method was reported in 2013 [22] to fabricate high density bulks and over 5 T has been achieved in a single 20 mm diameter bulk fabricated by hot-pressing ball-milled Mg and B powders [23]. Such approaches result in high J_c values, but large pressure apparatus are required, adding a significant practical limitation and reducing the feasibly-achievable sample dimensions [24]. Several techniques have also been proposed at ambient pressure, such as the capsule method [25] and Mg-liquid infiltration growth technique [24].

Significant improvements continue to be made in terms of in-field J_c and trapped field capability (fields given are surface fields, unless otherwise stated):

- 5.4 T at 12 K was achieved with a single 20 mm diameter MgB_2 bulk fabricated by hot-pressing ball-milled Mg and B powders [23];
- 3.75 T at 5 K with a pair of 32 mm diameter C-doped MgB_2 bulks fabricated by the infiltration and growth process [26];
- 3.72 T at 5 K with a 30 mm diameter MgB_2 bulk fabricated using ball-milled powders and a pressureless, powder-in-closed-tube synthesis technique [27];
- 3.6 T at 13.2 K and 2.8 T at 20 K with a single 38 mm diameter Ti-doped MgB_2 bulk [28] and 4.6 T at 14.1 K at the centre of a pair of such bulks [29], fabricated by the hot isostatic pressing (HIP) method;
- 3.14 T at 17.5 K with a pair of 25 mm diameter MgB_2 bulks fabricated by uniaxial hot pressing [30];
- 2.4 T at 15.9 K with a 30 mm diameter MgB_2 bulk fabricated using infiltration-capsule method [31];
- 2.8 T at 11 K with a pair of 30 mm diameter MgB_2 bulks fabricated by conventional sintering and 4.02 T at 11 K and 2.9 T at 20 K at the centre of the pair [32];
- 1.5 T at 16.4 K with a 30 mm diameter MgB_2 bulk fabricated by a capsule method [25]; and
- 1.3 T at 15 K with a 55 mm diameter sample (with a 6 mm central hole) fabricated by the reactive Mg-liquid infiltration (Mg-RLI) technique [33].

In [34], the techniques presented in this book are applied to investigate the magnetisation properties of MgB_2 bulks fabricated by different processes, including the influence of the geometric dimensions of the sample on the trapped field, and this is described in detail in section 3.1.1.3.

1.3.3 Bulk iron–pnictide superconductors

The discovery of superconductivity in iron–pnictide compounds in 2008 [35] raised the prospect of finding high-temperature superconductivity in materials other than cuprates and these materials exhibit a number of properties attractive for applications, including low anisotropy, high upper critical magnetic fields (H_{c_2}) in excess of 90 T and intrinsic critical current densities above 1 MA cm^{-2} (0 T, 4.2 K). It was shown recently in [36] that bulk iron–pnictide superconducting magnets capable of trapping over 1 T (5 K) and 0.5 T (20 K) can be fabricated with fine-grain polycrystalline $Ba_{0.6}K_{0.4}Fe_2As_2$ (Ba122). These Ba122 magnets were processed by a scalable, versatile and low-cost method using common industrial ceramic processing techniques. To fabricate the bulk Ba122 samples investigated in [36], Ba, K, Fe, and As in a molar ratio of 0.6:0.42:2:2 were reacted together by a mechanochemical reaction, followed by sintering in a hot isostatic press (HIP) at 600 °C using the procedure described in [37]. After bulk synthesis and subsequent re-milling, approximately 3–5 g of Ba122 powder was pressed into 15.9 mm diameter pellets and then further densified in a cold isostatic press (CIP) at 276 MPa. These were then wrapped with Ag foil and inserted into a steel tube that was carefully machined to the diameter of the pellet and foil. Both ends of the steel tube were welded shut under vacuum using chamfered plugs that helped the steel tube to compress the pellet. After welding, the tubes were swaged and cold isostatically pressed to further shape and densify them, reducing the diameter of the samples approximately 10%. Finally, the samples were sintered for 10 h at 600 °C in the HIP. After heat treatment, the steel tubes were sliced with a diamond saw to reveal the pellet surfaces.

In [38], the techniques presented in this eBook are applied to investigate the magnetisation properties of Ba122 bulks (singularly and multiple samples stacked together), including the influence of the geometric dimensions of the sample on the trapped field, and this is described in detail in section 3.1.1.4.

1.4 Magnetisation of bulk superconductors

In order to utilise a bulk superconductor as a TFM, they must firstly be magnetised by applying and removing a large magnetic field. There are three magnetisation techniques for magnetising a bulk superconductor that are in common use: zero-field-cooled (ZFC) and field-cooled (FC) magnetisation, as well as pulsed field magnetisation (PFM). The basic principles behind the ZFC and FC techniques are shown in figure 1.2 using Bean's 'critical state' model [39, 40], which was developed

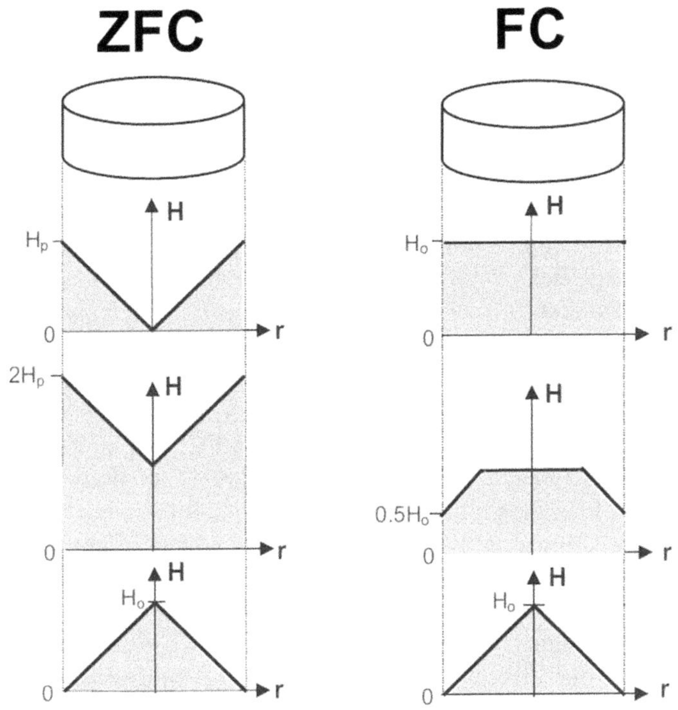

Figure 1.2. Zero-field-cooled (ZFC) and field-cooled (FC) bulk superconductor magnetisation techniques using the Bean model [39, 40] approximation [41]. For ZFC (left), a magnetic field equal to the full penetration field, H_p, is applied to a pre-cooled sample (top panel), which is increased to $2H_p$ (middle panel), and then the field is removed, trapping a field $H_0 = H_p$ (bottom panel). For FC (right), a magnetic field H_0 is applied to a sample above its critical temperature T_c (top panel), which is then cooled below T_c and the field is slowly removed (middle panel; where the applied field is reduced to $0.5H_0$), trapping a field equal to the applied field H_0 (bottom panel). Reproduced with permission from [41].

in the 1960s to explain macroscopically the irreversible magnetisation behaviour of hard type II superconductors. The underlying assumptions for the Bean model are

a) the relative permeability of type II superconductor, μ_r, is 1 in the mixed state, as in free space, i.e., $\boldsymbol{B} = \mu_0 \boldsymbol{H}$;
b) the applied magnetic field is assumed such that $H > H_{c_1}$ (ignoring the Meissner phase below H_{c_1}, which is assumed to be negligible) where the field begins to penetrate partially into the material in the form of vortices and $H < H_{c_2}$, H_{irr};
c) any electromotive force in the region penetrated by external flux lines, however small, will induce the critical current density, $J = J_c$, to flow in the superconductor and the current remains zero where the field has not penetrated ($J = 0$); and
d) there is no field dependence of the critical current density.

Based on the above assumptions, Ampere's law can be written as ($\nabla \times \boldsymbol{B} = \mu_0 \boldsymbol{J}_c$) and for an infinitely long superconducting slab, where a is the half-width of the slab, the supercurrent flowing is the gradient of the magnetic field penetrating the superconductor. The full penetration field, H_p, in figure 1.2 is then $H_p = J_c a$. In chapter 2, this will be related to the calculation of the peak trapped magnetic flux density of a superconducting disc of finite thickness.

In ZFC magnetisation, the superconductor is cooled below its critical temperature, T_c, prior to the application of a large magnetic field, typically of several tesla or more. A magnetic field of comparatively lower magnitude is applied in the FC process to a superconductor at a temperature above T_c, which is then cooled below T_c before the magnetising field is removed. In either case, to trap the maximum possible field corresponding to the sample's flux trapping ability, i.e., a supercurrent J_c flowing through the entire cross-section of the sample, the magnitude of the applied field needs to be at least H_p (at least H_p in FC; $2H_p$ in ZFC).

The ZFC and FC magnetisation techniques are often used to determine the maximum trapped field capability of a bulk superconductor, but invariably require large magnetising coils, usually superconducting, which is impractical for most engineering applications of these materials. The PFM technique is similar to ZFC, except that the large magnetic field is applied via a pulse on the order of milliseconds, rather than ramped up and down slowly over a period of many minutes or even hours. Achieving reliable, *in situ* magnetisation is crucial to producing a competitive and compact machine design in trapped flux-type rotating machines, for example, and the PFM technique shows great promise as a compact, mobile and relatively inexpensive magnetisation technique for the magnetisation of these and other TFM-based devices.

1.4.1 Pulsed field magnetisation

Although PFM has a number of important advantages over the ZFC and FC techniques, one significant and challenging issue with the PFM technique is that the trapped field produced is generally smaller than that of ZFC or FC, particularly at lower operating temperatures, due to the large temperature rise ΔT associated with the rapid dynamic movement of the magnetic flux vortices within the superconductor during the PFM process [42]. However, at higher operating temperatures (closer to T_c, such as 77 K) fields have been trapped close to that of FC values [16, 43–46]. The record trapped field produced by PFM to date is only 5.2 T at 29 K [47], which is significantly less than the record trapped field (17.6 T at 26 K [1]), magnetised by FC. It should also be noted that the 5.2 T was recorded on the surface of the sample and 17.6 T at the centre of a stack of two samples. Nevertheless, this is still around one third of the trapped field capability of bulk HTS materials.

Therefore, it is important to understand the flux dynamics when magnetising a bulk superconductor using the PFM technique to achieve an optimum trapped field profile and to investigate how these results translate into a practical magnetising technique for applications such as trapped flux-type superconducting electric machines [48]. There are many experimental results that have been published in the literature and the following issues have been confirmed for the PFM technique:

- The temperature rise increases with increasing magnitude of the applied pulsed magnetic field and with decreasing operating temperature for an identically applied field, which comes from the stronger magnetic forces and the decreased heat capacity of the bulk at lower temperatures, respectively [49];
- The temperature rise during PFM increases for increased J_c values of the bulk, which is strongly related to the pinning force, F_p [50];
- The temperature rise, which arises due to the pinning loss and the viscous loss [50, 51] takes place adiabatically during PFM, because a large amount of the heat generation takes place instantaneously and the cooling power has a finite limitation. The low thermal conductivity of the bulk also seriously affects the thermal and electromagnetic responses; and
- The experimental setup, including the thermal pathways between the cooling system (either cryogenic fluid or cryocooler) and the sample and the type and form of the epoxy impregnation of the bulk (if present), can heavily influence the resultant magnetisation [49].

All of these results can be clarified by measuring the maximum temperature, T_{max}, of the bulk surface after the application of the pulsed field. Because of significant temperature rises and the inverse relationship between J_c and the temperature T, the bulk cannot be magnetised to its highest potential, but also the bulk is not irreversibly damaged by PFM, as far as has been observed thus far.

The PFM technique, in particular, has many design considerations: the magnitude and duration of the pulse(s), the number of applied pulses, and the type and shape of the magnetising coil/fixture. The temperature(s) at which the pulse(s) are applied is also crucial, as is the method of cooling, which can be done by submersion in a liquid cryogen or via a cryocooler/cold head. All of these considerations can be analysed through numerical modelling in order to comprehensively optimise the PFM setup and such models can be useful in a number of other ways: to interpret experimental results and understand the physical properties of the material during the magnetisation process, as well as propose and analyse new magnetising techniques and assess the performance of practical bulk superconductor-based applications, which are more difficult to do experimentally. The subsequent chapters in this book will provide readers with numerical analysis techniques to model the magnetisation of bulk superconductors based on the finite element method (FEM), such as the coupled electromagnetic-thermal models required to analyse PFM.

1.5 Bulk superconductor applications

Bulk superconducting materials can be used in applications in several ways: flux pinning, flux trapping (i.e., as a TFM), flux shielding, magnetic field lensing and as a conductor alternative.

1.5.1 Flux pinning applications

1.5.1.1 Levitation

Bulk superconductors can act like a strong diamagnet and the stability of the resulting levitation is greatly enhanced by the flux pinning ability of type II superconductors. Hence, a PM can stably levitate above a bulk superconductor, even when turned on its side or upside down (with the PM suspended below the bulk). In general, the forces generated by the Meissner effect are less than those generated by flux pinning and are ignored in most analyses and the equilibrium position of the levitated PM and the forces exerted on it are dependent on whether the superconductor is cooled in close proximity to or far away from the PM, based on similar considerations to those given in section 1.4 related to ZFC and FC magnetisation processes.

A powerful demonstration of bulk superconductor levitation is the levitation of large masses, including the levitation of a sumo wrestler in Japan of around 200 kg. Using this PM-bulk HTS interaction, masses over 1000 kg have been levitated [11]. This has a number of potential applications, including:

- Levitation of cryogenic structures (e.g., cryogenic storage tanks, transmission lines) to reduce heat leak, and one of the earliest studies focussed on the levitation of hydrogen storage tanks for automobiles [11];
- Conveyor systems in the manufacturing industry and clean rooms, where stringent requirements mandate no mechanical contact [11]; and
- Magnetically levitated (MAGLEV) transportation that allows vehicles to move with minimal friction where the diamagnetic bulks are placed over a PM guideway. A prototype 'vehicle' with seats for two people, 'SuperTrans,' has been developed in Germany [52] that runs above an 80 m indoor circular track and a 12 passenger per compartment design was developed in Brazil [53] and tested on a 100 m outdoor track. There are also a number of groups in China actively working on levitation for MAGLEV technology [54, 55], including evacuated tube transport [56]. Additionally, there is a possibility of exploiting the flux trapping capability of bulk superconductors to act as more powerful PMs that would allow stronger levitation forces and higher levitation heights.

1.5.1.2 Magnetic bearings, flywheel energy storage and superconducting mixers

A cylindrically symmetrical PM-bulk HTS arrangement would allow free rotation about its axis of symmetry to form a passively stable, wear-free and unlubricated bearing with a very low coefficient of friction (rotational drag force with respect to the bearing load), orders of magnitude smaller than mechanical bearings. A number of bulk HTS bearings have been developed for various applications with good success [57, 58].

Several major projects have focussed on using this phenomenon in flywheel energy storage systems [58], where kinetic energy is stored in a large rotating mass, to reduce rotational losses and obtain upwards of 90% storage efficiency, including a 5 kWh/100 kW Boeing flywheel energy storage system [59], a NEDO 10 kWh/100 kW flywheel project [60], and a 5 kWh/250 kW ATZ-magnet-motor (MM) flywheel system [61]. The equivalent technology using superconducting coils would require coils of hundreds of metres in diameter [57].

Noncontact, superconducting mixers using the flux pinning property of bulk superconductors have also been proposed [62], where PM-bulk HTS magnetic coupling creates a support mechanism similar to that of a thrust-type mechanical bearing, which allows contact-free mixing.

1.5.2 Flux trapping applications

1.5.2.1 Magnetic separation

The force exerted on a magnetic particle can be given by [63]:

$$F = AB\frac{\mathrm{d}B}{\mathrm{d}z} \tag{1.2}$$

where A is a constant, B is the magnitude of the magnetic field and $\mathrm{d}B/\mathrm{d}z$ is the magnetic field gradient. In addition to the possibility of providing a large B, the magnetic field gradients, $\mathrm{d}B/\mathrm{d}z$ and $\mathrm{d}B/\mathrm{d}r$, are naturally large in a magnetised bulk superconductor. Hence, there has been interest in using these materials in magnetic separation applications. An obvious and useful application is the purification of water (waste or otherwise), as well as capturing useful materials from a slurry or other fluid of interest. A number of successful examples exist in the literature, including the separation of fine α-hematite (Fe_2O_3) particles from slurry [64, 65], removing ferromagnetic precipitates containing Mn from waste water [66] and recovery of Ni compounds from plating waste [67]. An array of five bulk superconductors, magnetised by PFM, has also been successfully used to separate ferromagnetic particles (magnetite, Fe_3O_4) and paramagnetic ones (α-hematite) dispersed in water [68, 69].

1.5.2.2 Rotating machines

Electric motors and the systems they drive are the single largest electricity end-use, accounting for 43%–46% of global electricity consumption and about 6 Gt of CO_2 emissions annually [70]. An electric machine converts mechanical to electrical energy, or vice versa, by the interaction of current-carrying conductors and magnetic fields. The output power, P, for any electric machine is directly proportional to geometrical factors that are mainly topology dependent, the magnetic air-gap flux density, B, the current loading, A, and the mechanical rotation speed, f_{mech} [71]:

$$P = \frac{1}{1 + K_\Phi}\frac{m}{m_1}\frac{\pi}{2}K_e K_i K_p \lambda_0^2 B A f_{\mathrm{mech}} D^2 L \eta \tag{1.3}$$

The current loading, A, is either thermally limited or with advanced cooling technology by the efficiency required by the particular application. In order to increase the output power, one could increase the size of the machine (diameter, D, and length, L), which obviously makes the machine larger and heavier, or the rotation speed, f_{mech}. Bulk superconductors can potentially provide magnetic flux densities an order of magnitude higher than conventional PMs from a much smaller volume of material than wire-wound coils, and are therefore highly attractive for next-generation electric machines to increase *B* [72]. Research efforts on bulk superconductor-based machines have been on a smaller scale compared with their wire-wound counterparts, but there have been a number of successful demonstrations of the technology, as well as various novel conceptual designs proposed [72], including:

- An axial gap-type superconducting synchronous machine developed by Izumi *et al* [73] in Japan for electric ship propulsion. Pulsed current excitation of copper stator windings is used to magnetise the GdBCO bulk high-temperature superconductors that act as the magnetic field poles in the rotor. These field poles are cooled by a free-convection thermosyphon using Gifford–McMahon (GM) cryocoolers with neon gas as the cryogen;
- A radial gap-type, all-superconducting machine developed by Coombs *et al* [74] in the UK that used bulk HTS materials stuck to the outside of a cylindrical rotor and HTS coils in the armature windings. The bulks are magnetised *ex situ* using a separate arrangement of copper magnetising coils;
- As part of a research project sponsored by NASA and the US Department of Defense, the University Research Engineering and Technology Institute (URETI) on Aeropropulsion and Power investigated an exotic machine design comprised of Bi-2223 coils and bulk YBCO plates acting as TFMs [75]. A two-stage cooling system is used to exploit the difference in the T_cs of the two superconducting materials to magnetise the bulk plates using FC magnetisation [75].

The unique properties of bulk superconductors have also been exploited in reluctance (flux shielding) and hysteresis (flux pinning) motors [76]. A primary advantage of such motors is that it is not necessary to magnetise the bulk as per trapped flux-type motors. The torque of a reluctance motor is proportional to the difference in flux in the direct ('easy' magnetic path) and quadrature ('difficult' magnetic path) axes arising from the interleaving of magnetic and non-magnetic parts. The magnetic field along the direct axis is enhanced by using the shielding (diamagnetic) property of the bulk superconductor, reinforcing the low reluctance path provided by the ferromagnetic material in the rotor [76–78]. However, such motors only provide an increase in torque of about one third that of conventional ones and the use of ferromagnetic materials limits the flux density to less than 2 T due to saturation and the air-gap field can be up to 1 T (but usually less). These motors can also show hysteresis motor-like behaviour due to the flux pinning property of the bulk material. A conventional hysteresis motor has a ferromagnetic rotor and torque is produced by the interaction of the stator ('driving') field and magnetisation of the rotor by this field. Similar to an induction motor, there is a 'slippage' (lag) between the driving field and the rotor. This lag is independent of the rotational speed, so a constant torque can be obtained from start-up to the synchronous speed. Several prototypes, using cylinders, rings or discs of bulk superconducting materials, have been designed and tested [79–82], which exploit the hysteresis of a type II superconductor arising from flux pinning [76]. The main magnetic field must be produced by the stator windings and these motors tend to have a low operating power factor, efficiency and torque density compared with other types of motor.

1.5.2.3 Portable NMR/MRI systems

Ring-shaped bulk superconductors, in which a probe is inserted into the central bore, can provide reasonably uniform, high magnetic fields that can be used to improve the performance of NMR and MRI systems and allow for a compact and cryogen-free design. In 2007, the first proton NMR signal at 2.89 T (123 MHz) was reported using such a system, which consisted of two SmBCO bulk rings with an outer diameter of 36 mm and inner diameter of 7 mm [83] with an NMR spectrum linewidth of around 228 kHz (1850 ppm). A new magnet was developed using six annular EuBCO bulks (60 mm outer diameter, 16 mm inner diameter) stacked to a height of 70 mm and in 2011, the 1H spectra of toluene with a full width at half maximum of 0.4 ppm (80 Hz) was achieved in the 10 mm room-temperature bore [84]. The first magnetic resonance images using annular EuBCO bulks were also reported in 2011, with a magnetic homogeneity of 37 ppm (peak-to-peak) achieved in a Φ 6.2 mm × 9.1 mm cylindrical space with first order shimming [85]. Using this system, a clear 3D image of a chemically-fixed mouse foetus was acquired.

1.5.2.4 Lorentz force velocimetry

Velocity or flow rate measurement of fluids is of great importance to a wide range of industries and Lorentz force velocimetry (LFV) provides a noncontact electromagnetic flow measurement technique that is particularly attractive to measure opaque, hot and/or aggressive fluids, such as liquid metals, glass melts or molten salts. The basic operating principle of LFV is that when an electrically conductive fluid or object moves through a transverse magnetic field, eddy currents are induced in the fluid/object according to Faraday's law and the interaction between the magnetic field and eddy currents creates a Lorentz force that acts to brake the fluid/object. As a consequence of Newton's third law, a force, equal and opposite to the braking force, acts on the magnet system. The theory of LFV is detailed in [86] and can be expressed by

$$F_{\mathrm{L}} \sim \sigma v B^2 V \tag{1.4}$$

where the Lorentz force, F_{L}, depends linearly on the electrical conductivity, σ, mean flow velocity, v, magnetic flux density, B, squared and the characteristic volume, V. LFV systems have been tested successfully using conventional PMs and work well for liquid metals due to their high electrical conductance; however, such systems are limited, even using an optimised Halbach array [87], because of the limited surface magnetic flux density available from PMs. It is expected that the use of bulk superconductors, capable of providing several tesla, would significantly enhance LFV performance, as well as enable LFV even in weakly-conducting and slow-moving fluids, such as glass melts [88].

1.5.2.5 Other applications

Using the high trapped field capability of bulk superconductors, the Moses effect (where the free surface of water is deformed by a magnetic field of several tesla) has been observed, as well as magnetic levitation of diamagnetic materials and magnetic orientation of organic polymers and gels and carbon nanotubes, which is particularly attractive for improving crystal growth of organic semiconductors and other materials. Bulk superconductors have also been used successfully in an MDDS system [89], where a drug bound to small, biocompatible magnetic particles is directed to a specific target region in the body using the force exerted on the particles by the bulk magnet's field, which is (as for magnetic separation, see section 1.5.2.1) dependent on the magnitude of the magnetic field, as well as its gradient.

1.5.3 Flux shielding applications

In addition to acting as TFMs with high magnetic fields, hollow, cylindrical or tube-shaped bulk superconductors are also able to shield magnetic fields in a more efficient way than an equivalent ferromagnetic cylinder [90–92]. In [91], the authors present an analysis of the magnetic shielding properties of a polycrystalline Bi-2223 superconducting tube subject to AC and DC axial magnetic fields and the tube is shown to effectively shield an axial field up to a threshold axial field, B_{lim}, that is determined by the length of the tube and its mean radius, but also the exact $J_c(B)$ dependence of the superconductor. For decreasing tube length, B_{lim} can be severely reduced due to demagnetisation effects. Two mechanisms for the field penetration in the superconductor are described: one from the external surface and one from the opening ends of the tube; the latter being suppressed for long tubes. Finally, the shielding factor is increased for increasing frequency. In [92], this study is extended to analyse the effect of a cap on the ends of the tube, effectively closing the tube, and for tubes of shorter length, this has a noticeable, positive effect on the maximum shieldable field and the uniformity of the shielding. In addition, the effect of a non-superconducting joint and of inhomogeneities in the superconducting properties is investigated, which can both act to reduce the magnitude and location of maximum field attenuation. Finally, in [93], the authors investigate the shielding properties of hybrid ferromagnetic-superconductor shielding tubes, which had previously been studied mostly experimentally [94–96]. The addition of a ferromagnetic cylinder either inside or outside the superconducting cylinder can substantially improve the shielding limit, which is maximised when the ferromagnetic tube is placed outside the superconducting tube. The shielding effect in this configuration is decomposed into two processes: attenuation from the ferromagnetic layer (independent of the superconducting layer) and a decrease in magnetic flux density in the wall of the superconducting tube, according to its own shielding factor. More information on shielding applications of bulk superconductors and modelling thereof is provided in section 4.2.3.

1.5.4 Magnetic lens

The shielding effect can also be exploited to concentrate a magnetic field, higher than the externally applied magnetic field, in the bore of a magnetic lens fabricated using ring-shaped bulk superconductors. The important features of such a lens are a tapered inner diameter and slits to suppress the circumferential shielding current that shields the magnetic flux from the ring's interior [97]. A concentrated field of 12.42 T has been achieved at 20 K for a background field of 8 T using a bulk GdBCO magnetic lens [98] and 30.4 T has been achieved at the centre of the lens in a background field of 28.3 T [99]. Using a bulk MgB_2 magnetic lens, a concentrated field of 2.18 T at 4.2 K has also been achieved for a background field of 1 T [97]. However, since the magnetic lens effect vanishes after the external field is removed, this external magnet must be operated continuously, which consumes a large amount of energy. In [100], a new concept of a hybrid trapped field magnet lens (HTFML) was proposed, which combines the bulk magnetic lens with a cylindrical bulk TFM (exploiting the vortex pinning effect). The HTFML can reliably generate a magnetic field at the centre of the lens higher than the trapped field in the cylindrical bulk TFM and the external magnetising field, even after the external field is removed. Such a device could prove useful to enhance the magnetic field in the bore of a bulk superconducting NMR/MRI system to improve its resolution [100]. A summary of this device, including the numerical modelling framework and results, is provided in section 4.2.4.

1.5.5 Conductor alternative

There has also been interest in using the high current-carrying capability of bulk superconducting materials in superconducting current leads to supply currents to a superconducting device from an external, room-temperature power supply in order to reduce heat leak [101]. In addition, the highly nonlinear resistivity of the superconductor and transition between the superconducting and normal state can be used in a device known as a superconducting fault current limiter (SFCL) to limit fault currents in power systems. A bar-shaped bulk material can be used for this purpose [102], but one difficulty faced is the uniformity of the transition across the entire material due to an inhomogeneous distribution of J_c. If, during the fault, only part of the bulk transitions to the resistive state, local hot spots can develop that can cause irreversible damage to the material and this so-called quench has a low propagation speed because of the low thermal conductivity of HTS materials [102]. A novel type of bulk magnet has also been proposed [103], where a thin, spiral-shaped coil-like element is machined from a large single-grain bulk and current can be applied via leads at each end of the spiral.

In the next chapter, some of the analytical and numerical modelling techniques that can be used to analyse the magnetisation properties of bulk superconductors are introduced, including an overview of the general process used to model such materials using FEM. An overview of modelling the electrical and thermal properties of bulk superconducting materials, including how to couple these together, is also provided.

References

[1] Durrell J H *et al* 2014 *Supercond. Sci. Technol.* **27** 082001

[2] Tomita M and Murakami M 2003 *Nature* **421** 517–20

[3] Bardeen J, Cooper L N and Schrieffer J R 1957 *Phys. Rev.* **106** 162–4
[4] Lee P J 2001 *Engineering Superconductivity* (New York: Wiley)
[5] Cardwell D A 1998 *Mater. Sci. Eng. B* **53** 1–10
[6] Matsushita T 2007 *Flux Pinning in Superconductors* (Berlin, Heidelberg: Springer)
[7] Anderson P W and Kim Y B 1964 *Rev. Mod. Phys.* **36** 39–43
[8] Yeshurun Y, Malozemoff A P and Shaulov A 1996 *Rev. Mod. Phys.* **68** 911–49
[9] Zeldov E *et al* 1989 *Phys. Rev. Lett.* **62** 3093–6
[10] Zeldov E *et al* 1990 *Appl. Phys. Lett.* **56** 680–2
[11] Hull J R and Murakami M 2004 *Proc. IEEE* **92** 1705–18
[12] Nariki S, Sakai N and Murakami M 2005 *Supercond. Sci. Technol.* **18** S126–30
[13] Fuchs G *et al* 2003 *J. Low Temp. Phys.* **133** 159–79
[14] Weinstein R, Sawh R, Ren Y and Parks D 1998 *Mater. Sci. Eng. B* **53** 38–44
[15] Sawh R-P, Weinstein R, Carpenter K, Parks D and Davey K 2013 *Supercond. Sci. Technol.* **26** 105014
[16] Weinstein R, Parks D, Sawh R and Davey K 2015 *IEEE Trans. Appl. Supercond.* **25** 6601106
[17] Robinowitz M, Arrowsmith H W and Dahlgren S D 1977 *Appl. Phys. Lett.* **30** 607–9
[18] Nagamatsu J *et al* 2011 *Nature* **410** 63–4
[19] Xu M *et al* 2001 *Appl. Phys. Lett.* **79** 2779–81
[20] Larbalestier D C *et al* 2001 *Nature* **410** 186–9
[21] Kambara M *et al* 2001 *Supercond. Sci. Technol.* **14** L5
[22] Sasaki T, Naito T and Fujishiro H 2013 *Phys. Procedia* **45** 93–6
[23] Fuchs G *et al* 2013 *Supercond. Sci. Technol.* **26** 122002
[24] Bhagurkar A G *et al* 2015 *Supercond. Sci. Technol.* **28** 015012
[25] Naito T, Sasaki T and Fujishiro H 2012 *Supercond. Sci. Technol.* **25** 095012
[26] Bhagurkar A G, Yamamoto A, Wang L, Xia M, Dennis A R, Durrell J H, Aljohani T A, Babu N H and Cardwell D A 2018 *Sci. Rep.* **8** 13320
[27] Sugino S, Yamamoto A, Shimoyama J and Kishio K 2015 *Supercond. Sci. Technol.* **28** 055016
[28] Yoshida T, Naito T and Fujishiro H 2015 *IEEE Trans. Appl. Supercond.* **25** 6801204
[29] Naito T, Yoshida T and Fujishiro H 2015 *Supercond. Sci. Technol.* **28** 095009
[30] Durrell J H *et al* 2012 *Supercond. Sci. Technol.* **25** 112002
[31] Naito T *et al* 2016 *Supercond. Sci. Technol.* **29** 115003
[32] Yamamoto A *et al* 2014 *Appl. Phys. Lett.* **105** 032601
[33] Perini E *et al* 2011 *IEEE Trans. Appl. Supercond.* **21** 2690–3
[34] Zou J *et al* 2015 *Supercond. Sci. Technol.* **28** 075009
[35] Kamihara Y, Watanabe T, Hirano M and Hosono H 2008 *J. Am. Chem. Soc.* **130** 3296–7
[36] Weiss J D *et al* 2015 *Supercond. Sci. Technol.* **28** 112001
[37] Weiss J D, Jiang J, Polyanskii A A and Hellstrom E E 2013 *Supercond. Sci. Technol.* **26** 074003
[38] Ainslie M D, Yamamoto A, Fujishiro H, Weiss J D and Hellstrom E E 2017 *Supercond. Sci. Technol.* **30** 105009
[39] Bean C P 1962 *Phys. Rev. Lett.* **8** 250
[40] Bean C P 1962 *Rev. Mod. Phys.* **36** 31
[41] Krabbes G, Fuchs G, Canders W-R, May H and Palka R 2006 *High Temperature Superconductor Bulk Materials* (Weinheim: Wiley) p 106

[42] Fujishiro H *et al* 2009 *IEEE Trans. Appl. Supercond.* **19** 3545–8
[43] Itoh Y and Mizutani U 1996 Japan *J. Appl. Phys.* **35** 2114–25
[44] Itoh Y, Yanagi Y and Mizutani U 1997 *J. Appl. Phys.* **82** 5600–11
[45] Ikuta H, Ishihara H, Yanagi Y, Itoh Y and Mizutani U 2002 *Supercond. Sci. Technol.* **15** 606–12
[46] Mizutani U *et al* 1998 *Appl. Supercond.* **6** 235–46
[47] Fujishiro H, Tateiwa T, Fujiwara A, Oka T and Hayashi H 2006 *Physica* C **445–48** 334–8
[48] Ainslie M D *et al* 2014 *Supercond. Sci. Technol.* **27** 065008
[49] Fujishiro H, Oka T, Yokoyama K and Noto K 2003 *Supercond. Sci. Technol.* **16** 809–14
[50] Fujishiro H *et al* 2005 *Supercond. Sci. Technol.* **18** 158–65
[51] Fujishiro H, Kaneyama M, Tateiwa T and Oka T 2006 *J. Phys.: Conf. Ser.* **43** 405–8
[52] Schultz L *et al* 2005 *IEEE Trans. Appl. Supercond.* **15** 2301–5
[53] Sotelo G G *et al* 2010 *J. Phys.: Conf. Ser.* **234** 032054
[54] Wang J *et al* 2011 *IEEE Trans. Appl. Supercond.* **21** 1551–5
[55] Deng Z, He D and Zheng J 2015 *IEEE Trans. Appl. Supercond.* **25** 3600106
[56] Jiang J, Bai X, Wu L and Zhang Y 2012 *J. Mod. Transp.* **20** 108–14
[57] Hull J R 2000 *Supercond. Sci. Technol.* **13** R1
[58] Werfel F N *et al* 2012 *Supercond. Sci. Technol.* **25** 014007
[59] Strasik M *et al* 2010 *Supercond. Sci. Technol.* **23** 2010
[60] Koshizuka N 2010 The superconducting magnetic bearings and magnetic clutches for flywheel energy storage *7th Int. Workshop on Processing and Applications of Superconducting (RE)BCO Large Grain Materials (PASREG 2010) (Washington, DC)*
[61] Werfel F N *et al* 2010 *J. Phys.: Conf. Ser.* **234** 032062
[62] Koyama F, Akiyama S and Murakami M 2006 *Supercond. Sci. Technol.* **19** 572–74
[63] Murakami M 2007 *Int. J. Appl. Ceram. Technol.* **4** 225–41
[64] Yokoyama K *et al* 2003 *IEEE Trans. Appl. Supercond.* **13** 1592–5
[65] Hayashi H, Nagabuchi N, Saho N, Mizumori T and Asano K 2004 *Physica* C **412–14** 766–70
[66] Oka T *et al* 2009 *J. Phys.: Conf. Ser.* **156** 012031
[67] Oka T *et al* 2016 *IEEE Trans. Appl. Supercond.* **26** 3700204
[68] Hiyama T *et al* 2008 *Physica* C **468** 1469–72
[69] Fujishiro H, Miura T, Naito T and Hayashi H 2010 *J. Phys.: Conf. Ser.* **234** 032015
[70] Waide P and Brunner C U 2011 Energy-efficiency policy opportunities for electric motor-driven systems, international energy agency doi: 10.1787/5kgg52gb9gjd-en
[71] Huang S, Luo J, Leonardi F and Lipo T A 1998 *IEEE Trans. Ind. Appl.* **34** 92–7
[72] Zhou D *et al* 2012 *Supercond. Sci. Technol.* **25** 103001
[73] Zhang Y, Zhou D, Ida T, Miki M and Izumi M 2016 *Supercond. Sci. Technol.* **29** 044005
[74] Jiang Y *et al* 2008 *Supercond. Sci. Technol.* **21** 065001
[75] Luongo C A *et al* 2009 *IEEE Trans. Appl. Supercond.* **19** 1055–68
[76] Barnes G J 2000 *Supercond. Sci. Technol.* **13** 875–78
[77] Kovalev L *et al* 2002 *Supercond. Sci. Technol.* **15** 817–22
[78] Oswald B *et al* 1999 *IEEE Trans. Appl. Supercond.* **9** 1201–4
[79] Kovalev L K *et al* 1998 *Mater. Sci. Eng. B* **53** 216–9
[80] Muta I, Jung H, Nakamura T and Hoshino T 2002 *Physica* C **372–376** 1531–4
[81] Nakamura T and Jung H J 2006 *Physica* C **445–448** 1115–8
[82] Tsuboi Y and Ohsaki H 2003 *IEEE Trans. Appl. Supercond.* **13** 2210–3

[83] Nakamura T, Itoh Y, Yoshikawa M, Oka T and Uzawa J 2007 *Concepts. Magn. Reson. B* **31B-2** 65–70
[84] Nakamura T *et al* 2011 *TEION KOGAKU (J. Cryogen. Supercond. Soc. Japan)* **46** 139–48
[85] Ogawa K, Nakamura T, Terada Y, Kose K and Haishi T 2011 *Appl. Phys. Lett.* **98** 234101
[86] Thess A, Votyakov E V and Kolesnikov Y 2006 *Phys. Rev. Lett.* **96** 164501
[87] Halbedel B, Resagk C, Wegfrass A, Diethold C, Werner M, Hilbrunner F and Thess A 2013 *Flow, Turbul. Combust.* **92** 361–9
[88] Vakaliuk O V, Ainslie M D and Halbedel B 2018 *Supercond. Sci. Technol.* **31** 084003
[89] Nakagawa K, Mishima F, Akiyama Y and Nishijima S 2012 *IEEE Trans. Appl. Supercond.* **22** 4903804
[90] Gyoda M, Torii S and Ebihara D 1998 *Adv. Supercond. XII* (Tokyo: Springer), 1353–6
[91] Denis S *et al* 2007 *Supercond. Sci. Technol.* **20** 192–201
[92] Denis S, Dirickx M, Vanderbemden P, Ausloos M and Vanderheyden B 2007 *Supercond. Sci. Technol.* **20** 418–27
[93] Lousberg G P *et al* 2010 *IEEE Trans. Appl. Supercond.* **20** 33–41
[94] Itoh M, Ohyama T, Hoshino K, Ishigaki H and Minemoto T 1993 *IEEE Trans. Appl. Supercond.* **3** 181–4
[95] Itoh M, Mori K and Minemoto T 1996 *IEEE Trans. Magn.* **32** 2605–8
[96] Mori K, Minemoto T and Itoh M 1997 *IEEE Trans. Appl. Supercond.* **7** 378–81
[97] Zhang Z Y, Choi S, Matsumoto S, Teranishi R, Giunchi G, Figini Albisetti A and Kiyoshi T 2012 *Supercond. Sci. Technol.* **25** 025009
[98] Zhang Z Y, Matsumoto S, Teranishi R and Kiyoshi T 2012 *Supercond. Sci. Technol.* **25** 115012
[99] Choi S, Yoon J-H, Lee B-S, Won M-S, Ok J-W, Zhang Z-Y, Kiyoshi T, Matsumoto S and Lee S-H 2012 *J. Appl. Phys.* **111** 07E728
[100] Takahashi K, Fujishiro H and Ainslie M D 2018 *Supercond. Sci. Technol.* **31** 044005
[101] Fujishiro H, Ikebe M, Teshima H and Hirano H 2006 *IEEE Trans. Appl. Supercond.* **16** 1007–10
[102] Tixador P *et al* 2000 *Supercond. Sci. Technol.* **13** 493–7
[103] Morita M, Kumada M, Sato A, Teshima H and Hirano H 2007 *Physica* C **463–465** 387–93

Chapter 2

Numerical modelling of bulk superconducting materials

Chapter 2 introduces some of the analytical and numerical modelling techniques that can be used to analyse the magnetisation properties of bulk superconductors and provides an overview of the general process used to model such materials using the finite element method. An overview of modelling the electrical and thermal properties of bulk superconducting materials, including how to couple these together, is provided. This is particularly important as the superconductor's electrical properties vary with temperature and under certain magnetisation processes and conditions, the bulk can experience significant changes in temperature.

2.1 Modelling of bulk superconductors

There exist various analytical and numerical models that have been proposed for simulating the properties of bulk superconducting materials, which will be outlined in this chapter. Numerical modelling is a powerful and cost-effective tool to understand the physical mechanisms of bulk superconductor magnetisation, to assist in interpretation of experimental results, to propose and analyse new magnetisation techniques, to guide superconducting material processing and practical device design, and to predict the performance of bulk superconductor-based devices.

Analytical models can be easier and faster to implement in comparison to numerical methods, such as those based on the finite element method (FEM), but are limited to specific, simplified geometries and assumptions [1]. Numerical models can overcome some of these limitations and simulate more complex geometries and situations, but this comes at the expense of increased computational requirements, more complex software implementation and longer computational times [2]. An example is the multi-physical, time-dependent numerical model required for the pulsed field magnetisation of a bulk superconductor, which requires both electromagnetic and thermal considerations that can change rapidly over a very short time scale on the order of milliseconds [1].

doi:10.1088/978-0-7503-1332-2ch2

2.1.1 Analytical techniques

Based on the Bean's critical state model [3, 4] (see section 1.4) and application of the Biot–Savart law, the peak trapped magnetic flux density, B_t, at the centre of the top surface of a single-grain bulk superconductor, oriented with its thickness parallel to the *c*-axis, due to an induced, persistent supercurrent, is given in its simplest form by

$$B_t = k\mu_0 J_c a \tag{2.1}$$

where μ_0 is the permeability of free space, J_c is the critical current density of the superconducting material, and a is the sample radius. k is a correction factor to the simple Bean (infinite slab) approximation due to the finite thickness, t, of a disc-shaped bulk superconductor sample, given by [5, 6]:

$$k = \frac{t}{2a} \ln\left(\frac{a}{t} + \sqrt{1 + \left(\frac{a}{t}\right)^2}\right) \tag{2.2}$$

A more general formula for the magnetic flux density at any height, z, above a superconducting disc of radius a and thickness t was presented in [6–9]:

$$B_t(z) = \frac{\mu_0 J_c}{2}\left[(z + t)\ln\left(\frac{a + \sqrt{a^2 + (z + t)^2}}{z + t}\right) - z \cdot \ln\left(\frac{a + \sqrt{a^2 + z^2}}{z}\right)\right] \tag{2.3}$$

Figure 2.1 shows a comparison of the analytical results for the peak, central trapped magnetic flux density on the top surface ($z = 0$ mm) of a bulk superconductor using equation (2.1) and the magnetic flux density at various heights between $z = 0$ mm to $z = +2$ mm above the top surface of the bulk using equation (2.3). The results are also compared with the trapped magnetic flux density profile across the top surface of the bulk using an FEM model that models a disc carrying a constant current density equal to the critical current density, J_c (labelled 'FEM CL'). The assumed parameters are: radius, $a = 5$ mm; thickness, $t = 10$ mm; $J_c = 1 \times 10^8$ A m^{-2}. The top surface is along the $z = 0$ mm line and the bulk is centred at $r = 0$ mm.

COMSOL Multiphysics® software implementation:

The example model is built in COMSOL Multiphysics® version 5.4 (www.comsol.com. COMSOL AB, Stockholm, Sweden). Additional modules required: AC/DC.

(Example 1—FEM CL.mph is available at http://iopscience.iop.org/book/978-0-7503-1332-2)

If you do not have a license for the COMSOL Multiphysics® software, you can contact your COMSOL® representative at www.comsol.com/contact to request a free trial license.

Equation (2.1) implies two main approaches for improving the field trapping ability of a bulk superconductor: (1) by enhancing the critical current density through improved flux pinning and increased sample homogeneity, and (2) by increasing the sample size. It is generally accepted that an aspect ratio of between 1 and 1.5 (radius/thickness) is the best compromise between the accessible, surface trapped field and volume of superconducting material [10–12], taking into account

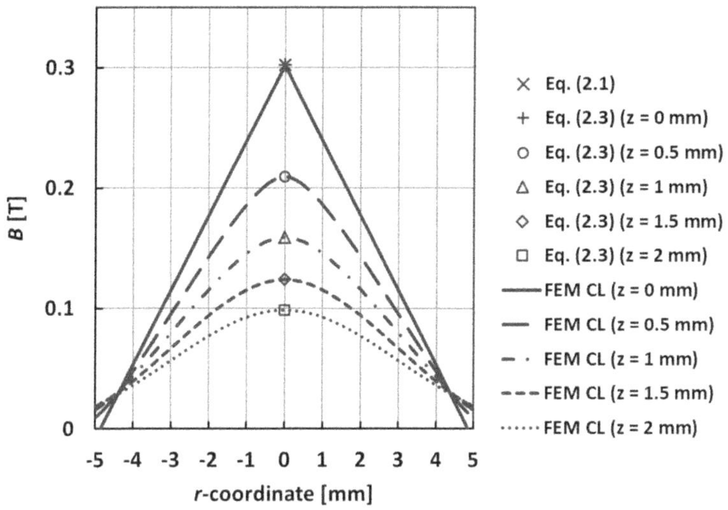

Figure 2.1. Analytical results (symbols) for (a) the peak, central trapped magnetic flux density on the top surface (z = 0 mm) of a bulk superconductor using equation (2.1) based on Bean's critical state model and application of the Biot–Savart law, and (b) using equation (2.3) for the magnetic flux density at various heights between z = 0 mm to z = +2 mm above the top surface of the bulk. These results are compared with the trapped magnetic flux density profile across the top surface of the bulk using a finite element method (FEM) model that models a disc carrying a constant current density equal to the critical current density, J_c (labelled 'FEM CL'). The assumed parameters are: radius, a = 5 mm; thickness, t = 10 mm; $J_c = 1 \times 10^8$ A m^{-2}. The top surface is along the z = 0 mm line and the bulk is centred at r = 0 mm.

the correction factor k and the $J_c(B)$ characteristics of practical materials. In addition to these factors, the performance of a bulk superconductor can be limited by strong Lorentz forces that result in large mechanical stresses on the material that can induce cracks in the sample and severely degrade the trapped field [13], and a heat exchange rate with the cryogenic fluid or cooling apparatus that is too low can result in a significant temperature rise in the material, especially during its magnetisation process [14].

A comprehensive overview of analytical methods and formulae for modelling HTS materials is presented in [15], but this work focuses chiefly on thin strips and stacks thereof, and is therefore most applicable to HTS wires, cables and related devices. Analytical expressions have been derived for the current and field distributions using the Bean assumption (constant J_c; see section 1.4) for an infinitely thin disc for the static nonlinear case [16], as well as for perpendicular time-varying periodic external magnetic fields [17]. Numerical solutions also exist for superconducting discs of finite thickness [18–21], and in [22], Brandt calculated the electromagnetic response of a disk of finite thickness in a perpendicular time-varying magnetic field, extending his work on a thin strip in [23]. Such methods are usually based on the critical state model (constant J_c, no frequency dependence and no flux creep) and can only deal with simple, uniform fields. Numerical methods, which are introduced in the following section, can be used model more complex, practical situations, for arbitrary geometries and applied fields, and where the superconductor's behaviour can vary depending on a number of different variables.

2.1.2 Numerical techniques

There exist a number of different numerical techniques that have been applied variously to the modelling of superconducting materials [2, 24], including the finite difference (FD) method [25], the combined sand-pile/Biot–Savart method [26–29], the Fourier transform method [30], minimisation or variational techniques [31–40] and FEM. In general, FEM techniques are the most commonly used and developed methods, and in this book, a versatile numerical modelling technique based on FEM is utilised to analyse various problems related to bulk superconductor magnetisation. However, interested readers are encouraged to consult the references listed above to learn more about other techniques.

2.2 Finite element method (FEM)

FEM is the most utilised method by far for solving and analysing many engineering problems, due to the relative ease with which different geometries, especially complicated ones, can be modelled. There also now exist numerous commercial software packages, such as COMSOL Multiphysics®, ANSYS® and FlexPDE®, which have reasonably gentle learning curves. The basic principle of FEM is to divide the physical system under investigation into subregions or subdomains consisting of numbers of discrete elements, which are simple units for which the physical behaviour can be readily analysed. The complete set of all discretised elements is known as the mesh.

The corresponding partial differential equations representing the underlying physical behaviour (electromagnetic, thermal and so on) are then approximated and solved within each of these small elements, which are interconnected at points common to two or more elements (nodes or nodal points), boundary lines and/or surfaces. The nodal values of the field variable(s) (variables of interest governed by the underlying equations) are explicitly calculated based on the given initial and boundary conditions, and since these are calculated locally, this results in a relatively accurate solution. These nodal values are then used to approximate values at non-nodal points (within the elements), determined using interpolation (or shape) functions (usually polynomial: linear, quadratic or higher order) that are based on the element type chosen. In FEM, weighted residuals are used, which means the numerical errors are weighted over the whole domain [41].

Mesh refinement is a key part of any FEM model and sometimes one must make a trade-off between accuracy and computational time. A finer mesh (more, smaller elements) results in a more accurate solution, nearer the 'true' solution, but the computational time and memory required increases, sometimes prohibitively. Appropriate meshing is also related to the geometry, as well as the nature of the physics of the system under analysis. The example models provided throughout this book include meshes that, in the authors' experience, provide good accuracy without taking too long to solve.

In the following section, the main steps required when modelling bulk superconductors using FEM are summarised.

2.2.1 Modelling bulk superconductors using FEM

2.2.1.1 Geometry, including magnetisation fixture

The geometry of the numerical model for a disc-shaped bulk superconductor can generally be simplified using a two-dimensional (2D) axisymmetrical model (r, z, Φ in a cylindrical coordinate system), assuming that its properties are homogeneous in the Φ-direction, i.e., around the bulk's *ab*-plane. However, a three-dimensional (3D) geometry (x-, y-, z-coordinates) is required for any inhomogeneous material properties around the *ab*-plane and for shapes without cylindrical symmetry, such as rectangular-shaped bulks. It can still be useful to make use of symmetry in a 3D model by applying appropriate boundary conditions to model, for example, one half or one quarter of the bulk, which can reduce the number of mesh elements required, and thus improve computational memory requirements and speed, while retaining the necessary 3D effect.

The magnetising fixture can be simulated using uniform electromagnetic boundary conditions to represent an externally applied magnetic field, which is usually applied to the bulk along the z-axis, perpendicular to its top surface (and parallel to the bulk's c-axis). It is also possible to introduce the magnetising coil itself as a separate entity in the model using, for example, a copper coil subdomain and setting an appropriate integral constraint for the current flowing in the coil. This is useful to investigate the influence of the type of magnetising coil and its shape on the magnetisation properties, e.g., solenoid [42, 43], vortex [44] and split [45, 46] coils. The appropriate electromagnetic boundary conditions for either of these options are detailed later in this chapter in section 2.2.1.3.3.

Figure 2.2 highlights two such 2D axisymmetric models: one where uniform electromagnetic boundary conditions are used to apply the external magnetic field and one where a current is applied to a copper coil subdomain representing the solenoid coil that provides the applied field. The specific boundary conditions given for H_r, H_z are related to the $\boldsymbol{H}$-formulation that will be introduced in more detail in the next section and throughout this book.

Additional subdomains can also be included for other materials, including a sample holder (e.g., copper, stainless steel) [46], impregnation materials (e.g., epoxy resin) [46] and ferromagnetic materials [46, 47].

Finally, if thermal effects are included in the model (see section 2.2.1.4), cooling can be simulated assuming submersion in a liquid cryogen [48], e.g., liquid nitrogen at 77 K, or a cold head and vacuum chamber [49].

2.2.1.2 Electromagnetic formulation

FEM-based techniques have been applied to many superconducting material problems using a variety of formulations [2, 49, 50]: namely the $\boldsymbol{A}$-V [51–53], $\boldsymbol{T}$-Ω [54–56], $\boldsymbol{H}$ [42, 48, 57–67] and $\boldsymbol{E}$ [49, 68–70] formulations, and Campbell's equation [53, 71]. Maxwell's equations can be written in each of these formulations and these

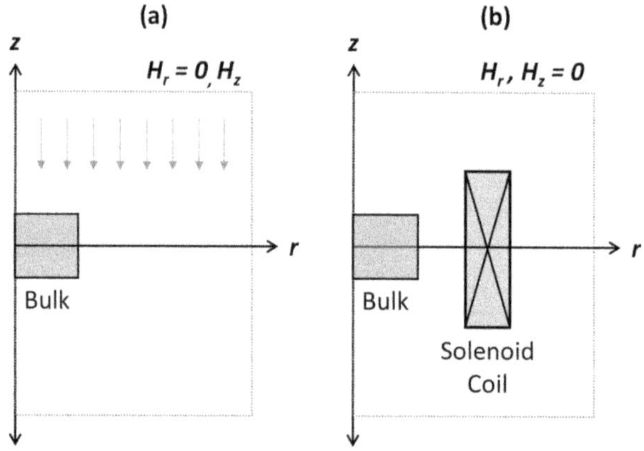

Figure 2.2. 2D axisymmetric models (r, z, Φ in a cylindrical coordinate system) of a disc-shaped bulk superconductor, which assumes its properties are homogeneous in the Φ-direction, i.e., around the bulk's *ab*-plane with: (a) uniform electromagnetic boundary conditions to represent an externally applied magnetic field, H_z, applied to the bulk along the z-axis, perpendicular to its top surface (and parallel to the bulk's c-axis), and (b) a copper coil subdomain representing a solenoid coil, where the field is applied by setting an appropriate integral constraint for the current flowing in the coil.

formulations are equivalent in principle, i.e., the choice of formulation should not result in a different physical meaning of the solution, but the solutions of the corresponding partial differential equations (PDEs) can be very different [72]. Importantly, the choice of numerical method does impact on the complexity and computational time required to solve the model. It should be noted that many of the examples given in the preceding references are applied to geometries other than bulk superconductor discs, but in the remainder of this book, the application of such numerical techniques related specifically to bulk superconductor magnetisation is highlighted in more detail.

2.2.1.2.1 ***H*-formulation**

The majority of the modelling examples presented here are simulated using the $\boldsymbol{H}$-formulation, where the independent variables are the components of the magnetic field strength, $\boldsymbol{H}$. The governing equations are derived from Maxwell's equations—namely, Ampere's (2.4) and Faraday's (2.5) laws:

$$\nabla \times \boldsymbol{H} = \boldsymbol{J} \tag{2.4}$$

$$\nabla \times \boldsymbol{E} + \frac{\mathrm{d}\boldsymbol{B}}{\mathrm{d}t} = \nabla \times \boldsymbol{E} + \frac{\mathrm{d}(\mu_0 \mu_r \boldsymbol{H})}{\mathrm{d}t} = 0 \tag{2.5}$$

In the 2D axisymmetric model, $\boldsymbol{H} = [H_r, H_z]$, $\boldsymbol{J} = [J_\Phi]$ represents the current density and $\boldsymbol{E} = [E_\phi]$ represents the electric field. In 3D, $\boldsymbol{H} = [H_x, H_y, H_z]$, $\boldsymbol{J} = [J_x, J_y, J_z]$ and $\boldsymbol{E} = [E_x, E_y, E_z]$. μ_0 is the permeability of free space, and for the superconducting and air sub-domains, the relative permeability can be assumed as simply $\mu_r = 1$.

COMSOL Multiphysics® software implementation:

The electromagnetic formulation above can be implemented in the standard COMSOL Multiphysics® software package using the 'General Form PDE' interface, for which the following equation is satisfied:

$$e_a\frac{\partial^2 \boldsymbol{u}}{\partial t^2} + d_a\frac{\partial \boldsymbol{u}}{\partial t} + \nabla \cdot \Gamma = f \tag{2.6}$$

where $\boldsymbol{u} = [H_r, H_z]$ or $\boldsymbol{u} = [H_x, H_y, H_z]$. Equations (2.4) and (2.5) can be re-written in this form by specifying the appropriate coefficients. The PDE interface has been traditionally used in this manner for a variety of superconducting problems; however, in recent versions of COMSOL® there has been a dedicated interface—the 'Magnetic Field Formulation (mfh)' interface in the AC/DC module—that is particularly suitable for time-dependent modelling of materials, such as superconductors, with highly nonlinear *E–J* characteristics (see section 2.2.1.3.2).

2.2.1.3 Electrical properties

2.2.1.3.1 Critical current density $J_c(B, T)$

The most important characteristic of a superconductor for any practical superconducting device is its critical current density, J_c, which determines the peak trapped magnetic flux density, B_t, of a bulk superconductor (see equation (2.1)). Hence, a crucial aspect of bulk superconductor modelling is providing accurate input data for the in-field J_c, including any anisotropy, at different temperatures, i.e., $J_c(\boldsymbol{B}, T)$, of the material(s) under investigation. If accurate $J_c(\boldsymbol{B}, T)$ data is input into the model, it is then possible to accurately predict the magnitude and shape of the trapped field profile and the flux dynamics of the magnetisation process.

The Bean model (see section 1.4), which assumes a constant J_c, is a useful approximation for a type II superconductor in its mixed state and can be used to compare a basic numerical model with analytical models (see figure 2.1). This is always a particularly useful check in the initial stages of developing a new model.

Kim and Anderson [73, 74] showed that the J_c of type II superconductors (e.g., HTS materials) exhibits a strong dependence on temperature, as well as local magnetic field, which can be expressed by the generalised empirical relationship:

$$J_c(B, T) = \frac{J_{c_0}(T)}{\left(1 + \frac{B}{B_0}\right)^{\alpha}} \tag{2.7}$$

where J_c is the local critical current density, B is the magnitude of the local magnetic flux density, B_0 and α are material-dependent constants that can vary with temperature and J_{c_0} is the magnitude of the critical current density when the local field is zero. J_{c_0} can be considered a constant value or as a temperature-dependent function $J_{c_0}(T)$ if thermal considerations are necessary (see section 2.2.1.4). The material-dependent constants B_0 and α can be estimated by the trapped field profile of the bulk [75] or, more typically, by measurements of the magnetic moment hysteresis

(m-H) loops of sub-specimens taken from the bulk (or a representative sample) at different temperatures using a superconducting quantum interference device (SQUID) magnetometer [5, 11, 27, 76].

Equation (2.7) is suitable for modelling iron–pnictide and some (RE)BCO bulk superconductors. However, some (RE)BCO materials exhibit a so-called 'fish tail' or 'peak' effect, where there is a 'second peak anomaly' seen in their magnetisation loop. An exponential fitting function was proposed by Jirsa *et al* to represent such behaviour [77]:

$$J_c(B) = J_{c_1} \exp\left(-\frac{B}{B_L}\right) + J_{c_2}\frac{B}{B_{\max}} \exp\left[\frac{1}{y}\left(1 - \left(\frac{B}{B_{\max}}\right)^y\right)\right] \tag{2.8}$$

where J_{c_1}, B_L, J_{c_2}, $B_{\max}$ and y are fitting parameters. Figure 2.3 shows a comparison of some typically-used $J_c(B)$ assumptions—namely, the Bean, Kim (2.7) and Jirsa (2.8) models—compared with experimental data measured from a small sample taken from a representative 15wt% Ag-containing GdBCO bulk superconductor, measured at 77 K.

Inhomogeneities occur during the growth process of (RE)BCO bulk superconductors, including an uneven distribution of pinning sites and the formation of growth section regions (GSRs) and growth sector boundaries (GSBs), which leads to a variation of J_c with position: generally a lower J_c is found at the edge with respect to the centre and at the bottom with respect to the top [48]. This variation can play a

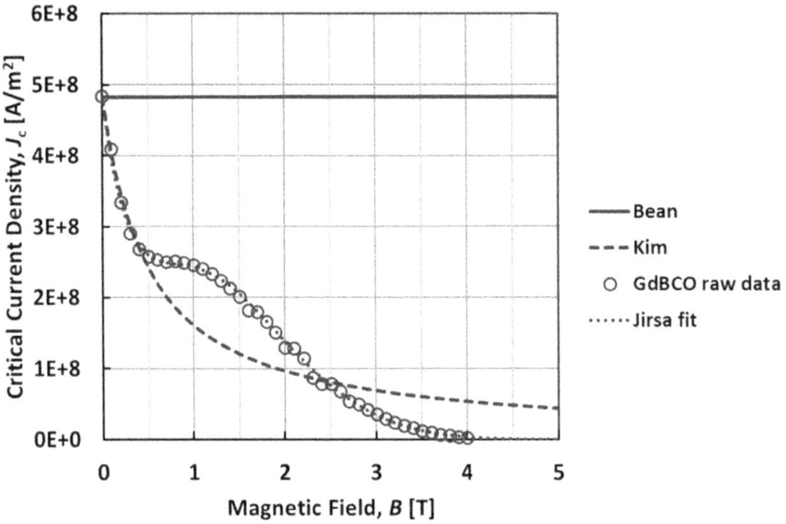

Figure 2.3. A comparison of different in-field critical current density, $J_c(B)$, assumptions: (a) Bean model (constant J_c), (b) Kim model (equation (2.7); $B_0 = 0.5$ T, $\alpha = 1$), (c) Experimental data measured from a small sample taken from a representative 15wt% Ag-containing GdBCO bulk superconductor (diameter 30 mm, thickness 15 mm), measured at 77 K [46], (d) Data fitting using the Jirsa equation (see (2.8); $J_{c_1} = 4.83 \times 10^8$ A m^{-2}, $B_L = 0.34$, $J_{c_2} = 2.2 \times 10^8$ A m^{-2}, $B_{\max} = 1.12$, $y = 1.87$). J_c for (a) and J_{c_0} for (b) corresponds to $J_c(0\text{ T}) = 4.83 \times 10^8$ A m^{-2} from (c).

significant role in the magnetisation of these materials, particularly for the PFM technique, where they can affect the dynamics of the magnetic flux penetration, the magnitude and shape of the trapped field and the maximum temperature rise, and different trapped field distributions can be observed in comparison to FC/ZFC magnetisation, particularly for lower applied fields. However, a position-dependent J_c is relatively easy to introduce into a model once the dependence on magnetic field, temperature and position is known.

For bulk MgB_2, a different equation is needed because there is no peak effect and there is a drastic reduction of J_c in the presence of a magnetic field. Typically, the following equation is used to represent bulk MgB_2 [78]:

$$J_c(B) = J_{c_0} \exp\left(-\left(\frac{B}{B_0}\right)^{\beta}\right) \tag{2.9}$$

where B_0 and β are material-dependent constants.

COMSOL Multiphysics® software implementation:

The $J_c(B, T)$ equations can be programmed directly in COMSOL® using user-defined variables, but the $J_c(B, T)$ data itself can also be input into the model using a two-variable direct interpolation [79, 80]. In the latter technique, the $J_c(B, T)$ data can be used directly and works like a look-up table. This avoids continually solving the $J_c(B, T)$ equation during the simulation, which can dramatically speed up the model, particularly when the fitting equation is complicated, e.g., [80], and improves accuracy.

2.2.1.3.2 E–J power law

It is assumed, in general, that the electric field, $\boldsymbol{E}$, is parallel to the current density, $\boldsymbol{J}$, such that $\boldsymbol{J} = \sigma\boldsymbol{E}$ or $\boldsymbol{E} = \rho\boldsymbol{J}$, where σ and ρ are the conductivity and resistivity, respectively. The highly nonlinear electrical resistivity of bulk superconducting materials is typically simulated using an E–J power law [81, 82], where E is proportional to J^n. In its simplest form,

$$E = E_0\left(\frac{J}{J_c}\right)^n \tag{2.10}$$

or in more detailed form,

$$\boldsymbol{E} = \frac{E_0}{J_c(B, T)}\left|\frac{J}{J_c(B, T)}\right|^{n-1}\boldsymbol{J} \tag{2.11}$$

Here, $E_0 = 1\ \mu\text{V cm}^{-1}$ is the characteristic electric field that defines J_c and n defines the steepness of the transition between the superconducting state and the normal state and accounts for flux creep relaxation, which results in a logarithmic decay of trapped field after the magnetising field is removed (see section 1.2.3). For bulk HTS materials, n usually varies between 5 (strong flux creep) and 50 (weak flux creep and the limiting value between HTS and LTS materials). For $n > 20$, it becomes a good approximation of the Bean model, for which $n \to \infty$. For this reason, many of the

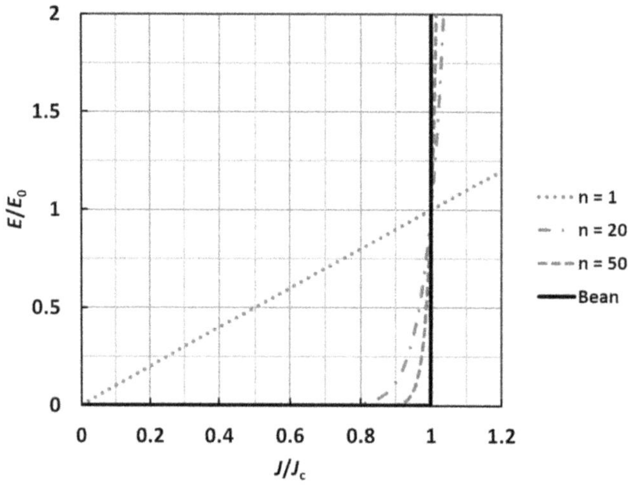

Figure 2.4. A comparison of the *E–J* power law (equation (2.10)), which is typically used to model the highly nonlinear electrical resistivity of bulk superconducting materials, for $n = 1$ (linear Ohm's law), 20 and 50, as well as the Bean model, for which $n \to \infty$.

models in the literature use an n value of around 20–30 as a reasonable value. Figure 2.4 shows a comparison of *E–J* curves based on equation (2.10) for n values of 1 (linear Ohm's law), 20 and 50, as well as the Bean model.

For a superconducting wire, the n value is usually estimated from experimentally-measured *I–V* curves; however, this is relatively uncommon, and reasonably difficult to measure (special current contacts are required to avoid the sample heating with large currents flowing), for bulk superconducting materials. As described in section 2.2.1.3.1 for J_c, one can also introduce an in-field, temperature-dependent n value should the relevant information be known [83].

2.2.1.3.3 Electromagnetic boundary conditions

In order to apply the external magnetising field to the bulk superconductor, appropriate electromagnetic boundary conditions must be set. The most straightforward option is to apply a uniform background magnetic field, typically applied parallel to the *c*-axis of the bulk superconductor, which is usually aligned along the *z*-axis in a model, such that the supercurrent flows around the *ab*-plane of the bulk. On the boundaries of the outer subdomain surrounding the bulk (see figure 2.2(a); usually assumed to be air), the magnetic field is set such that $H_r = 0$, $H_z = H_{app}(t) = f(t)$, which then applies a uniform external magnetic field, $f(t)$, to the bulk. One must ensure that this subdomain surrounding the bulk is suitably large enough such that the so-called self-field generated by the supercurrent flowing in the bulk is negligible at this outer boundary or errors will be induced in the solution [84]. Another option is to generate the external magnetising field via current flowing in a magnetising coil. In this case, a separate coil subdomain (usually copper) is introduced (see figure 2.2(b)) and a current is applied by making use of an integral constraint on this subdomain such that

$$I_{app}(t) = \int_s \boldsymbol{J} \cdot \mathrm{d}\boldsymbol{s} \tag{2.12}$$

where $I_{app}(t)$ is a function representing the current flowing in the magnetising coil and s is its cross-section. In this case, the magnetic field is set on the outer boundaries such that $H_r = H_z = 0$.

2.2.1.4 Thermal properties and electromagnetic-thermal coupling

In some cases—for example, for ZFC magnetisation (see section 1.4) where the applied field is ramped slowly at a fixed temperature, T—isothermal conditions can be assumed (i.e., a constant T) and hence no thermal model needs to be included. However, in cases where the bulk experiences a change in temperature, appropriate coupling of the electromagnetic model with a thermal model is required, including appropriate assumptions for the thermal properties. One such case is FC magnetisation (see section 1.4), where the magnetising field is applied to the bulk at a temperature above T_c, which is then cooled below T_c before the magnetising field is removed. It should also be noted that in this case, the electrical properties need careful consideration because the superconductor is initially in the normal state (above T_c) and transitions to the superconducting state when cooled below T_c. Another case is PFM, where there can be a large temperature rise ΔT associated with the rapid dynamic movement of the magnetic flux within the superconductor. Often the electromagnetic time constant (related to the pulsed field) is much smaller than the thermal time constant (related to the thermal conductivity and heat capacity of the bulk, as well as the efficacy of the cooling system employed), especially for (RE)BCO materials. This means that the thermal diffusion can be an order of magnitude slower than the magnetic flux diffusion: the temperature rise during PFM therefore essentially takes place adiabatically.

The preceding electromagnetic model can be coupled with a thermal model using the following thermal transient equation:

$$\rho \cdot C\frac{\mathrm{d}T}{\mathrm{d}t} = \nabla \cdot (\kappa \, \nabla T) + Q \tag{2.13}$$

where ρ is the mass density in kg m^{-3}, C is the specific heat in J kg^{-1} K^{-1}, κ is the thermal conductivity in W m^{-1} K^{-1} and Q is the heat source.

Suitable assumptions must be made for C and κ and figure 2.5 shows a comparison of typical measured values for C and κ for (RE)BCO and MgB_2 bulk superconductors. At the typical operating temperature, T_s, for each bulk ($T_s = 40$ K for (RE)BCO and $T_s = 15$ K for MgB_2), the specific heat per unit volume of bulk MgB_2 is three orders of magnitudes smaller and the thermal conductivity is about three times larger, compared with bulk (RE)BCO. What this means practically is that a larger temperature rise is expected for the MgB_2 for identical heat generation, but the generated heat diffuses faster. The reader can find more information on the thermal properties of bulk superconductors in appendix A.

If additional materials are included in the model, including, for example, a sample holder (e.g., copper, stainless steel), impregnation materials (e.g., epoxy resin) and ferromagnetic materials, then appropriate assumptions must also be made for these.

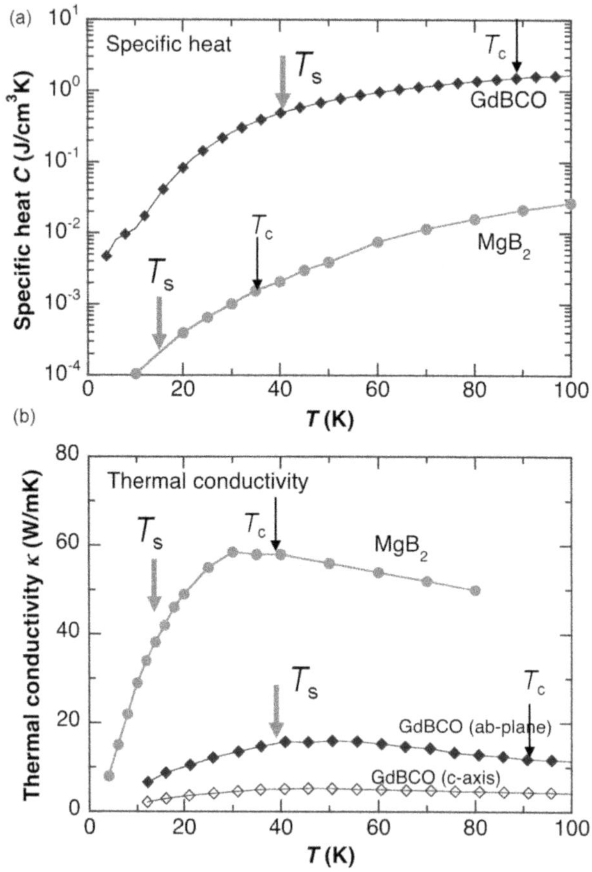

Figure 2.5. Comparison of (a) specific heat and (b) thermal conductivity for bulk (RE)BCO and MgB_2 superconductors. Reproduced by permission of IOP Publishing from [1].

If isothermal conditions are assumed while the applied field is ramped up or down, then $Q = 0$. To include any heat generated during the magnetisation process and couple the electromagnetic model to the thermal one, then $Q = E_\varphi \cdot J_\varphi$ (2D axisymmetric model) or $Q = E_{\text{norm}} \cdot J_{\text{norm}}$ (3D model), where $E_{\text{norm}} = \sqrt{E_x^2 + E_y^2 + E_z^2}$ and $J_{\text{norm}} = \sqrt{J_x^2 + J_y^2 + J_z^2}$.

Further, bidirectional coupling (electromagnetic <-> thermal) is then achieved through the assumed $J_c(B, T)$ characteristics by including the temperature dependence of J_c.

A simple approximation for the variation of J_c with temperature is to assume a linear relationship [83]:

$$J_c(B, T) = J_c(B)\frac{1 - \frac{T}{T_c}}{1 - \frac{T_0}{T_c}} \tag{2.14}$$

where T_0 is the reference temperature for some given $J_c(B)$ characteristics at $T = T_0$. One may also define a temperature-dependent n value, $n(T)$, in a similar way [83].

A more realistic assumption is to assume the temperature dependence is a function of $1 - (\frac{T}{T_c})^2$, such that

$$J_{c0}(T) = \alpha\left[1 - \left(\frac{T}{T_c}\right)^2\right]^m \tag{2.15}$$

where α is the critical current density extrapolated to $T = 0$ K and $m = 1.5$ [42, 48, 85, 86] or 2 [87, 88].

COMSOL Multiphysics® software implementation:

The thermal transient equation above (equation (2.13)) can be implemented in the standard COMSOL Multiphysics® software package using the 'Coefficient Form PDE' interface, for which the following equation is satisfied:

$$e_a\frac{\partial^2 \boldsymbol{u}}{\partial t^2} + d_a\frac{\partial \boldsymbol{u}}{\partial t} + \nabla \cdot (-c\nabla \boldsymbol{u} - \alpha \boldsymbol{u} + \gamma) + \beta \cdot \nabla \boldsymbol{u} + a\boldsymbol{u} = f \tag{2.16}$$

where $\boldsymbol{u} = [T]$. Equation (2.13) can be re-written in this form by specifying the appropriate coefficients. There is also a dedicated interface in COMSOL®: the 'Heat Transfer in Solids (ht)' interface in the Heat Transfer module.

The thermal properties of the equations can be programmed directly in COMSOL® as variables or the data itself (for example, see figure 2.5) can also be input into the model using a direct interpolation, like suggested for the $J_c(B, T)$ data in section 2.2.1.3.1. This also avoids continually solving additional equations during the simulation, which can speed up the model.

The next chapter details how the modelling frameworks described here can simulate the different magnetisation techniques for bulk superconductors—zero-field-cooled (ZFC) and field-cooled (FC) magnetisation, as well as pulsed field magnetisation (PFM)—and the influence of the electrical and thermal properties on magnetisation is also discussed.

References

[1] Ainslie M D and Fujishiro H 2015 *Supercond. Sci. Technol.* **28** 053002
[2] Grilli F *et al* 2014 *IEEE Trans. Appl. Supercond.* **24** 8200433
[3] Bean C P 1962 *Phys. Rev. Lett.* **8** 250
[4] Bean C P 1962 *Rev. Mod. Phys.* **36** 31
[5] Durrell J H *et al* 2012 *Supercond. Sci. Technol.* **25** 112002
[6] Naito T, Sasaki T and Fujishiro H 2012 *Supercond. Sci. Technol.* **25** 095012
[7] Cardwell D A *et al* 2005 *Supercond. Sci. Technol.* **18** S173–9
[8] Chen I-G, Liu J X, Weinstein R and Lau K 1992 *J. Appl. Phys.* **72** 1013–20
[9] Itoh Y and Mizutani U 1996 Japan *J. Appl. Phys.* **35** 2114–25
[10] Eisterer M *et al* 2006 *Supercond. Sci. Technol.* **19** S530–6
[11] Zou J *et al* 2015 *Supercond. Sci. Technol.* **28** 075009

[12] Ainslie M D *et al* 2017 *Supercond. Sci. Technol.* **30** 105009
[13] Durrell J H *et al* 2014 *Supercond. Sci. Technol.* **27** 082001
[14] Lousberg G *et al* 2009 *Supercond. Sci. Technol.* **22** 055005
[15] Mikitik G P, Mawatari Y, Wan A T S and Sirois F 2013 *IEEE Trans. Appl. Supercond.* **23** 8001920
[16] Mikheenko P N and Kuzolev Y E 1993 *Physica* C **204** 229–36
[17] Zhu J *et al* 1993 *Physica* C **212** 216–22
[18] Frankel D J 1979 *J. Appl. Phys.* **50** 5402
[19] Daeumling D and Larbalestier D C 1989 *Phys. Rev.* B **40** 9350
[20] Conner L W and Malozemoff A P 1991 *Phys. Rev.* B **43** 402
[21] Theuss H, Forkl A and Kronmüller H 1992 *Physica* C **190** 345
[22] Brandt E H 1994 *Phys. Rev.* B **50** 4034–50
[23] Brandt E H 1994 *Phys. Rev.* B **49** 9024–40
[24] Campbell A M 2011 *J. Supercond. Nov. Magn.* **24** 27–33
[25] Chechurin V L and Sarma M 1976 *IEEE Trans. Magn.* **6** 1048–9
[26] Fukai H, Tomita M, Murakami M and Nagatomo T 2001 *Physica* C **357** 774–6
[27] Fukai H, Tomita M, Murakami M and Nagatomo T 2002 *Supercond. Sci. Technol.* **15** 1054–7
[28] Aydiner A and Yanmaz E 2005 *Supercond. Sci. Technol.* **18** 1010–5
[29] Aydiner A and Yanmaz E 2010 *J. Supercond. Nov. Magn.* **23** 457–63
[30] Shen Z, Ainslie M D, Campbell A M and Cardwell D C 2015 *IEEE Trans. Appl. Supercond.* **25** 5201005
[31] Bossavit A 1994 *IEEE Trans. Magn.* **5** 3363–6
[32] Sanchez A and Navau C 2001 *Phys. Rev.* B **64** 214506
[33] Prigozhin L 1996 *J. Comput. Phys.* **129** 190–200
[34] Prigozhin L 1998 *J. Comput. Phys.* **144** 180–93
[35] Prigozhin L and Sokolovsky V 2011 *Supercond. Sci. Technol.* **24** 075012
[36] Pardo E, Gömöry F, Šouc J and Ceballos J M 2007 *Supercond. Sci. Technol.* **20** 351–64
[37] Pardo E 2008 *Supercond. Sci. Technol.* **21** 065014
[38] Kapolka M *et al* 2018 *IEEE Trans. Appl. Supercond.* **28** 6801405
[39] Kapolka M *et al* 2018 *IEEE Trans. Appl. Supercond.* **28** 8201206
[40] Kapolka M and Pardo E 2019 *Supercond. Sci. Technol.* **32** 054001
[41] Sirois F and Grilli F 2015 *Supercond. Sci. Technol.* **28** 043002
[42] Patel A and Glowacki B A 2012 *Supercond. Sci. Technol.* **25** 125015
[43] Shiraishi R and Ohsaki H 2006 *IEEE Trans. Appl. Supercond.* **16** 1794–7
[44] Fujishiro H, Naito T and Oyama M 2011 *Supercond. Sci. Technol.* **24** 075015
[45] Xu Z, Lewin R, Campbell A M, Cardwell D A and Jones H 2012 *Supercond. Sci. Technol.* **25** 025016
[46] Ainslie M D *et al* 2016 *Supercond. Sci. Technol.* **29** 074003
[47] Philippe M P *et al* 2015 *Supercond. Sci. Technol.* **28** 095008
[48] Ainslie M D *et al* 2014 *Supercond. Sci. Technol.* **27** 065008
[49] Vinot E, Meunier G and Tixador P 2000 *IEEE Trans. Magn.* **36** 1226–9
[50] Grilli F *et al* 2005 *IEEE Trans. Appl. Supercond.* **15** 17–25
[51] Stavrev S *et al* 2002 *IEEE Trans. Magn.* **38** 849–52
[52] Prigozhin L 1997 *IEEE Trans. Appl. Supercond.* **7** 3866–73
[53] Campbell A M 2009 *Supercond. Sci. Technol.* **22** 034005

[54] Amemiya N, Miyamoto K, Banno N and Tsukamoto O 1997 *IEEE Trans. Appl. Supercond.* **7** 2110–3
[55] Amemiya N *et al* 1998 *Physica* C **310** 30–5
[56] Banno N and Amemiya N 1999 *IEEE Trans. Appl. Supercond.* **9** 2561–4
[57] Pecher R, McCulloch M D, Chapman S J, Prigozhin L and Elliott C M 2003 *Proc. 6th EUCAS* pp 1–11
[58] Kajikawa K, Hayashi T, Yoshida R, Iwakuma M and Funaki K 2003 *IEEE Trans. Appl. Supercond.* **13** 3630
[59] Hong Z, Campbell A M and Coombs T A 2006 *Supercond. Sci. Technol.* **19** 1246
[60] Brambilla R, Grilli F and Martini L 2007 *Supercond. Sci. Technol.* **20** 16
[61] Sirois F, Dione M, Roy F, Grilli F and Dutoit B 2008 *J. Phys.: Conf. Ser.* **97** 012030
[62] Ainslie M D, Flack T J, Hong Z and Coombs T A 2011 *Int. J. Comput. Math. Electr. Electron. Eng.* **30** 762
[63] Ainslie M D *et al* 2011 *Supercond. Sci. Technol.* **24** 045005
[64] Ainslie M D, Flack T J and Campbell A M 2012 *Physica* C **472** 50
[65] Zhang M and Coombs T A 2012 *Supercond. Sci. Technol.* **25** 015009
[66] Ainslie M D, Yuan W and Flack T J 2013 *IEEE Trans. Appl. Supercond.* **23** 4700104
[67] Zermeno V, Grilli F and Sirois F 2013 *Supercond. Sci. Technol.* **26** 052001
[68] Sykulski J K, Stoll R L, Mahdi A E and Please C P 1997 *IEEE Trans. Magn.* **33** 1568–71
[69] Sykulski J K, Rotaru M and Stoll R L 1999 *Int. J. Comput. Math. Electr. Electron. Eng.* **18** 215–24
[70] Sykulski J K, Rotaru M and Stoll R L 2000 *IEEE Trans. Magn.* **36** 1178–82
[71] Campbell A M 2007 *Supercond. Sci. Technol.* **20** 292–5
[72] Grilli F, Brambilla R and Martini L 2007 *IEEE Trans. Appl. Supercond.* **17** 3155–8
[73] Kim Y B, Hampstead C F and Strnad A R 1963 *Phys. Rev. Lett.* **9** 306–9
[74] Anderson P W 1963 *Phys. Rev. Lett.* **9** 309–11
[75] Philippe M P *et al* 2014 *Physica* C **502** 20–30
[76] Fujishiro H *et al* 2011 *Supercond. Sci. Technol.* **24** 105003
[77] Jirsa M, Pust L, Dlouhy D and Koblischka M R 1997 *Phys. Rev.* B **55** 3276–84
[78] Xiang F X *et al* 2013 *Appl. Phys. Lett.* **102** 152601
[79] Hu D *et al* 2015 *Supercond. Sci. Technol.* **28** 065011
[80] Hu D, Ainslie M D, Raine M J, Hampshire D P and Zou J 2016 *IEEE Trans. Appl. Supercond.* **26** 6600906
[81] Plummer C J G and Evetts J E 1987 *IEEE Trans. Magn.* **23** 1179–82
[82] Rhyner J 1993 *Physica* C **212** 292–300
[83] Berger K, Leveque J, Netter D, Douine B and Rezzoug A 2007 *IEEE Trans. Appl. Supercond.* **17** 3028–31
[84] Queval L, Liu K, Yang W, Zermeno V M R and Ma G 2018 *Supercond. Sci. Technol.* **31** 084001
[85] Fujishiro H and Naito T 2010 *Supercond. Sci. Technol.* **23** 105021
[86] Ohsaki H, Matsumura S, Kawamoto S and Shiraishi R 2008 *AIP Conf. Proc.* **985** 999–1006
[87] Matsushita T *et al* 1993 *IEEE Trans. Appl. Supercond.* **3** 1045–8
[88] Ohsaki H, Shimosaki T and Nozawa N 2002 *Supercond. Sci. Technol.* **15** 754–8

Chapter 3

Modelling magnetisation of bulk superconductors

Chapter 3 details how the modelling frameworks described in chapter 2 can simulate the different magnetisation techniques for bulk superconductors: zero-field-cooled (ZFC) and field-cooled (FC) magnetisation, which are typically used to determine the maximum trapped field capability of a bulk superconductor, and pulsed field magnetisation (PFM), which shows great promise as a compact, mobile and relatively inexpensive magnetisation technique for magnetising bulk superconductors in practical applications. The influence of the electrical and thermal properties on the magnetisation of bulk high-temperature superconductors is also discussed, as well as other bulk superconducting materials, such as MgB_2 and iron–pnictide materials.

3.1 Magnetisation of bulk superconductors

As detailed in chapter 1, bulk superconductors can be used, when cooled to cryogenic temperatures, as super-strength, stable permanent magnets capable of generating fields of several tesla. However, in order to utilise a bulk superconductor in this way, they must firstly be magnetised, which essentially involves the application and removal of a large magnetic field that induces a circulating supercurrent in the material that flows without resistance. In this chapter, the three magnetisation techniques for magnetising a bulk superconductor that are in common use are discussed: namely, zero-field-cooled (ZFC) and field-cooled (FC) magnetisation, as well as pulsed field magnetisation (PFM).

3.1.1 Zero-field-cooled (ZFC) and field-cooled (FC) magnetisation

Although slow and requiring large magnetising coils, which is impractical for most practical engineering applications, the ZFC and FC magnetisation techniques are often used to determine the maximum trapped field capability of a bulk

doi:10.1088/978-0-7503-1332-2ch3

superconductor and to establish a reliable baseline to compare with PFM results. These techniques are often also used after the fabrication process to confirm single-grain growth and assess the uniformity of the sample.

3.1.1.1 Simulation of ZFC magnetisation

Using the framework described in chapter 2, a time-dependent model for ZFC magnetisation is straightforward to setup. As a simple example, we can assume the same parameters as in section 2.1.1: radius, a = 5 mm; thickness, t = 10 mm; $J_c = 1 \times 10^8$ A m^{-2}. The top surface is again along the z = 0 mm line and the bulk is centred at r = 0 mm. To provide an adequate comparison with the results shown in figure 2.1, an n value of 100 is assumed for the E-J power law (equation (2.11)) to closely approximate Bean's critical state model, and a uniform magnetising field is applied in the z-direction using boundary conditions, such that $H_z(t)$ is ramped linearly to some maximum value, H_{max}, and then ramped back down to zero, i.e., a triangular wave function. In this example, $\mu_0 H_{max}$ = 1.3 T, which is at least twice the penetration field, estimated by $\mu_0 J_c a \approx 0.63$ T using the Bean model (see section 1.4). The magnetising field is applied with a slow ramp rate of 20 mT s^{-1} so that thermal effects can be ignored.

Figure 3.1 shows a comparison of the trapped magnetic flux density profile across the top surface of the bulk (z = 0 mm) for this time-dependent FEM model (labelled 'FEM ZFC'), simulating ZFC magnetisation, with the analytical result and the stationary FEM model of a current-carrying disc (see section 2.1.1). There is, as expected, excellent agreement between the different methods.

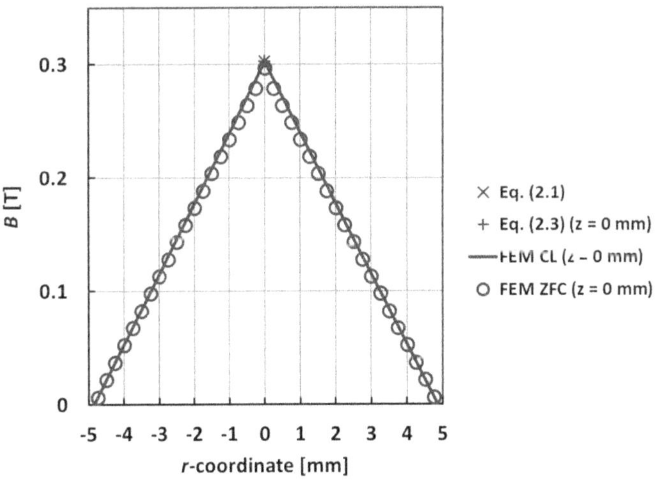

Figure 3.1. Trapped magnetic flux density profile across the top surface (z = 0 mm) of a bulk superconductor from a time-dependent model of zero-field-cooled (ZFC) magnetisation (indicated by circles and labelled 'FEM ZFC'). These results are compared with those shown previously in figure 2.1, section 2.1.1, for the analytical result using Bean's critical state model and the Biot–Savart law and the FEM model that models a disc carrying a constant current density equal to the critical current density, J_c ('FEM CL'). The assumed parameters are the same, i.e., radius, a = 5 mm; thickness, t = 10 mm; $J_c = 1 \times 10^8$ A m^{-2}.

Figure 3.2 shows (a) the flux penetration into the bulk (across the centre of the bulk, $z = -5$ mm) as the magnetising field is ramped up and (b) its subsequent exit from the bulk as the magnetising field is ramped back down to zero. One will observe that these results are consistent with those predicted by the Bean model (see figure 1.2, section 1.4). The peak of the applied field is reached at $t = 65$ s and it can be seen in figure 3.2(a) that the field at the centre of the bulk is greater than 0.63 T, the penetration field estimated using the Bean model. Thus, at the end of the ZFC process at $t = 130$ s, the bulk is fully magnetised.

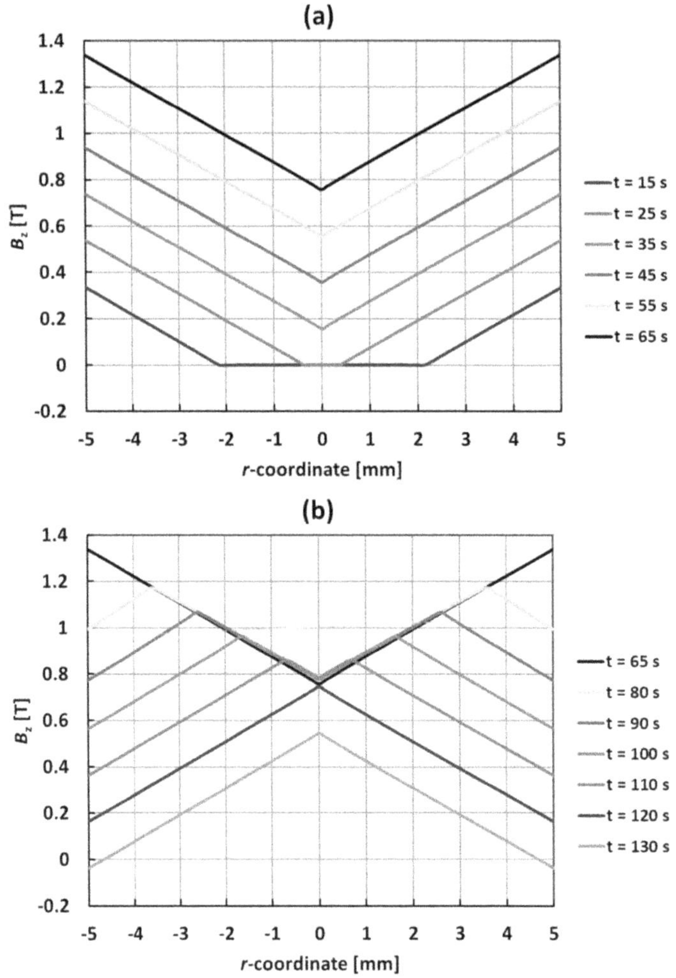

Figure 3.2. (a) Flux penetration into the bulk (across the centre of the bulk, $z = -5$ mm) as the magnetising field is ramped up ($t = 0$ s $\rightarrow$ 65 s) and (b) its subsequent exit from the bulk as the magnetising field is ramped back down to zero ($t = 65$ s $\rightarrow$ 130 s). The peak of the applied field is $\mu_0 H_{\max} = 1.3$ T, which is at least twice the penetration field, estimated by $\mu_0 J_c a \approx 0.63$ T using the Bean model. The magnetising field is applied with a slow ramp rate of 20 mT s^{-1} so that thermal effects can be ignored.

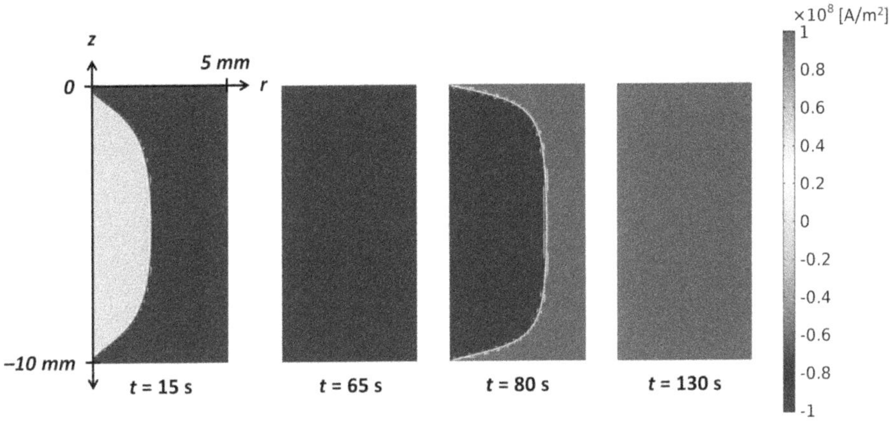

Figure 3.3. Current density, J_ϕ, distribution throughout the cross-section of the bulk during the ZFC magnetisation process at (a) t = 15 s, during the ramp-up of the magnetising field; (b) t = 65 s, at the peak of the magnetising field; (c) t = 80 s, during the ramp-down of the magnetising field; and (d) t = 130 s, at the end of the ZFC magnetisation process once the magnetising field has been ramped back down to zero. A current density $J_\varphi = J_c = 1 \times 10^8$ A m^{-2} flows where the flux has penetrated and the current remains zero ($J = 0$) where the flux has not penetrated, and this current density is equal to the gradient of the magnetic flux density (see figure 3.2).

Figure 3.3 shows the current density, J_ϕ, distribution throughout the cross-section of the bulk during the ZFC magnetisation process at (a) $t = 15$ s, during the ramp-up of the magnetising field; (b) $t = 65$ s, at the peak of the magnetising field; (c) $t = 80$ s, during the ramp-down of the magnetising field; and (d) $t = 130$ s, at the end of the ZFC magnetisation process once the magnetising field has been ramped back down to zero. A current density $J_\varphi = J_c = 1 \times 10^8$ A m^{-2} flows where the flux has penetrated and the current remains zero ($J = 0$) where the flux has not penetrated, and this current density is equal to the gradient of the magnetic flux density (figure 3.2). This gradient changes polarity as the flux exits the bulk, resulting in a supercurrent flowing in the opposite direction.

An astute reader will notice that the flux penetration is not uniform along the thickness of the bulk; the Bean model assumes an infinitely long slab with uniform flux penetration, but for a bulk of finite thickness, like in this example, the flux penetrates further into the bulk closer to the top and bottom surfaces, resulting in a deviation from simple analytical models. In fact, in [1], the authors show how the finite geometry of the bulk impacts on the dynamics of the flux penetration and highlight the usefulness of numerical simulations, which are used to clarify the misconception in [2] resulting from the use of a simple analytical analysis. The experimental results in [2] showed that, in the case of a partially-magnetised bulk magnetised by ZFC magnetisation, flux creep resulted in a reduced magnetic field measured at the centre of the top surface of the bulk. The simple, infinite slab Bean analysis in [2] suggested that magnetic flux vortices may move against the Lorentz force, contravening commonly accepted theory, but the numerical simulations in [1] show how the finite geometry of the bulk results in non-uniform flux penetration.

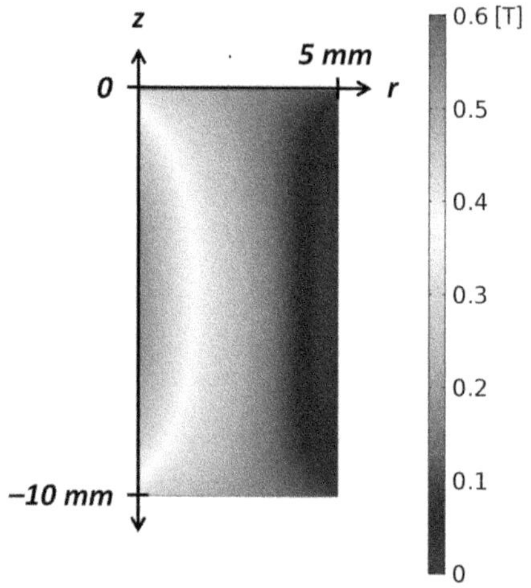

Figure 3.4. Final trapped magnetic flux density, $|\boldsymbol{B}|$, throughout the cross-section of the bulk at the end of the ZFC magnetisation process (t = 130 s).

This, in turn, results in an induced supercurrent that deviates from the simple Bean model and the Lorentz force acting on the vortices is no longer uniaxial.

Finally, figure 3.4 shows the final trapped magnetic flux density, $|\boldsymbol{B}|$, throughout the cross-section of the bulk at the end of the ZFC magnetisation process (t = 130 s).

COMSOL Multiphysics® software implementation:

The example model is built in COMSOL Multiphysics® version 5.4 (www.comsol.com. COMSOL AB, Stockholm, Sweden). Additional modules required: AC/DC.

(Example 2—FEM ZFC.mph is available at http://iopscience.iop.org/book/978-0-7503-1332-2)

If you do not have a license for the COMSOL Multiphysics® software, you can contact your COMSOL® representative at www.comsol.com/contact to request a free trial license.

3.1.1.2 Simulation of FC magnetisation

Setting up the simulation of FC magnetisation follows much the same process as for ZFC magnetisation, but there are a number of key differences. Since the temperature of the bulk changes significantly during FC magnetisation, the thermal model described in section 2.2.1.4 is required. It is assumed that the bulk is cooled via its bottom surface connected to a cryocooler of infinite cooling power with a perfect thermal connection. Thermal insulation boundary conditions are imposed on the side and top surfaces of the bulk, assuming a perfect vacuum. The assumed thermal properties (specific heat, C, and thermal conductivity, κ) for the bulk are those given in figure 2.5 in section 2.2.1.4; however, for simplicity, constant values of $\kappa_{ab} = \kappa_r$ = 20 W m^{-1} K^{-1} and $\kappa_c = \kappa_z$ = 4 W m^{-1} K^{-1} are assumed for the *ab*-plane and *c*-axis, respectively. The density of the bulk is assumed to be 5900 kg m^{-3} [3].

The electrical properties need careful consideration because the superconductor is initially in the normal state (above T_c) and transitions to the superconducting state when cooled below T_c (see section 2.2.1.4). The electrical resistivity is defined again by the E–J power law (equation (2.11)) with $n = 100$ for $T < T_c$ and $n = 1$ (linear resistivity) for $T \geqslant T_c$. The temperature-dependent $J_c(T)$ relationship given by equation (2.15) is assumed (see section 2.2.1.4) for $T < T_c$, with $J_c(77\text{ K}) = 1 \times 10^8$ A m^{-2}, $T_c = 92$ K and $m = 1.5$. α, the critical current density extrapolated to $T = 0$ K, is 6.1×10^8 A m^{-2}. For $T \geqslant T_c$, it is assumed that J_c has an extremely low value and is simply assumed to be equal to 1 A m^{-2}. Thus, J_c over the entire temperature range is assumed to be

$$J_{c0}(T) = 1 + 6.1 \times 10^8 \left[1 - \left(\frac{T}{92}\right)^2\right]^{1.5} \quad \text{for } T < T_c \tag{3.1}$$

and

$$J_{c0}(T) = 1 \quad \text{for } T \geqslant T_c \tag{3.2}$$

There are now three distinct steps in the numerical model: (1) ramping the external magnetising field up to its maximum magnitude, $\mu_0 H_{\text{max}}$, while the temperature of the bulk is held at $T > T_c$ ($T = 100$ K in this example); (2) cooling the bulk to an appropriate operating temperature, $T_{\text{op}} < T_c$, while the magnetising field is held at $\mu_0 H_{\text{max}}$; and (3) once the temperature has stabilised, ramping the magnetising field down to zero. The ramp rate of the magnetising field is the same as for the ZFC case (20 mT s^{-1}; see section 3.1.1.1) so that thermal effects can be ignored; hence, $Q = 0$ in equation (2.13), assuming isothermal conditions. $\mu_0 H_{\text{max}} = 0.65$ T, slightly higher than the full penetration field, 0.63 T. In this example: (1) $t = 0 \rightarrow 32.5$ s; (2) $t = 32.5 \rightarrow 250$ s; and (3) $t = 250 \rightarrow 282.5$ s.

Figure 3.5 shows a comparison of the trapped magnetic flux density profile across the top surface of the bulk ($z = 0$ mm) for this time-dependent FEM model (labelled 'FEM FC'), simulating FC magnetisation, with the results from section 2.1.1 and the time-dependent ZFC model above. There is again excellent agreement between the different methods.

Figure 3.6 shows the flux penetration and subsequent exit from the bulk (across the centre of the bulk, $z = -5$ mm) as the magnetising field is ramped down from $\mu_0 H_{\text{max}} = 0.65$ T, slightly higher than the estimated penetration field 0.63 T, at 20 mT s^{-1} ($t = 250$ s $\rightarrow$ 282.5 s). At the start of the ramp-down of the magnetising field ($t = 250$ s), the magnetic field within the bulk is equal to the external magnetising field (0.65 T). As this field is ramped down, the flux exits the bulk and establishes a magnetic field gradient equal to J_c (see figure 3.7). Finally, it can be seen at the end of the FC process, when the magnetising field reaches zero ($t = 282.5$ s), the bulk is fully magnetised in the same way as the ZFC process (see $t = 130$ s in figure 3.2(b)).

Figure 3.7 shows the current density, J_φ, distribution throughout the cross-section of the bulk during the FC magnetisation process at $t = 250$ s, at the start of the ramp-down of the magnetising field; $t = 256.5$ s and 269.5 s as this field is ramped down; and finally, at $t = 282.5$ s, at the end of the FC magnetisation process once the magnetising field has been ramped back down to zero. Initially, no current flows in the bulk and there is no magnetic field gradient ($t = 250$ s). The field gradient established as the flux exits the bulk (see figure 3.6) results in a current density

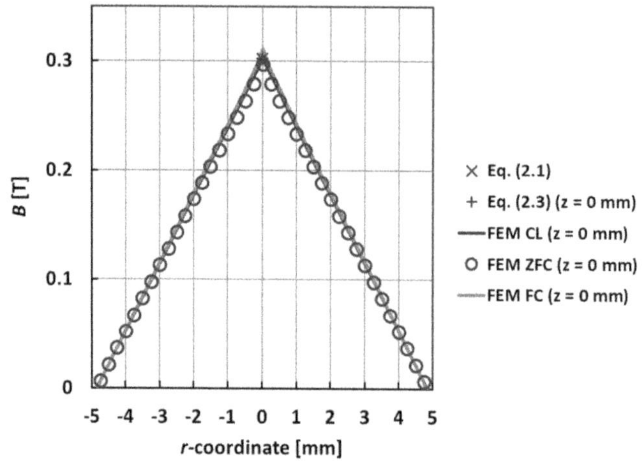

Figure 3.5. Trapped magnetic flux density profile across the top surface ($z = 0$ mm) of a bulk superconductor from a time-dependent model of field-cooled (FC) magnetisation (indicated by the red line and labelled 'FEM FC'). These results are compared with those shown previously in figure 2.1, section 2.1.1, for the analytical result using Bean's critical state model and the Biot–Savart law and the FEM model that models a disc carrying a constant current density equal to the critical current density, J_c ('FEM CL'), as well as the time-dependent ZFC magnetisation model presented in section 3.1.1.1 ('FEM ZFC'). The assumed parameters are the same, i.e., radius, $a = 5$ mm; thickness, $t = 10$ mm; $J_c = 1 \times 10^8$ A m^{-2}.

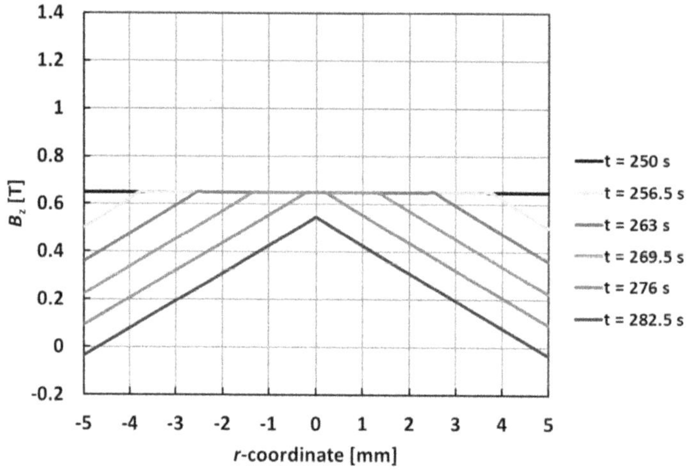

Figure 3.6. Flux penetration and subsequent exit from the bulk (across the centre of the bulk, $z = -5$ mm) as the magnetising is ramped down from $\mu_0 H_{max} = 0.65$ T, slightly higher than the penetration field 0.63 T, at 20 mT s^{-1} (start of ramp-down, $t = 250$ s → end of ramp-down, $t = 282.5$ s).

$J_\varphi = J_c = 1 \times 10^8$ A m^{-2} and the current remains zero ($J = 0$) where a field gradient is not yet established.

Figure 3.8 shows the final trapped magnetic flux density, $|\boldsymbol{B}|$, throughout the cross-section of the bulk at the end of the FC magnetisation process ($t = 282.5$ s). The end result is the same in terms of the trapped field and supercurrent flowing, but

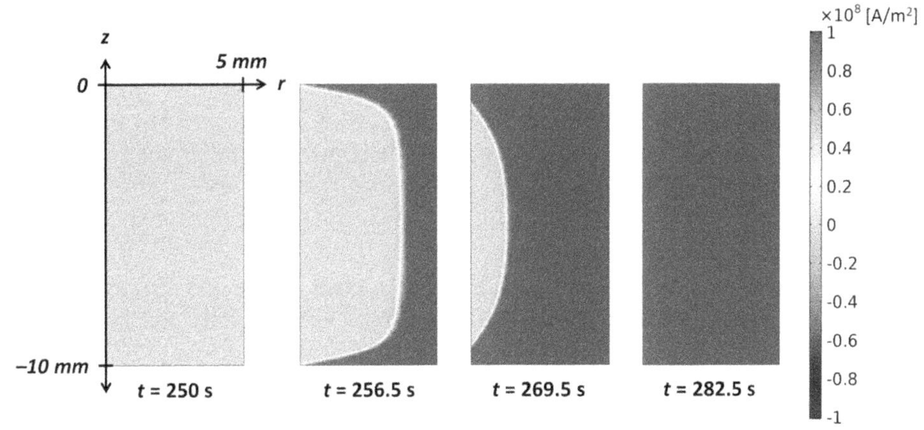

Figure 3.7. Current density, J_ϕ, distribution throughout the cross-section of the bulk during the FC magnetisation process at t = 250 s, at the start of the ramp-down of the magnetising field; t = 256.5 s and 269.5 s as this field is ramped down; and finally, at t = 282.5 s, at the end of the FC magnetisation process once the magnetising field has been ramped back down to zero. The magnetic field gradient established as the flux exits the bulk (see figure 3.6) results in a current density $J_\varphi = J_c = 1 \times 10^8$ A m^{-2} and the current remains zero (J = 0) where a field gradient is not yet established.

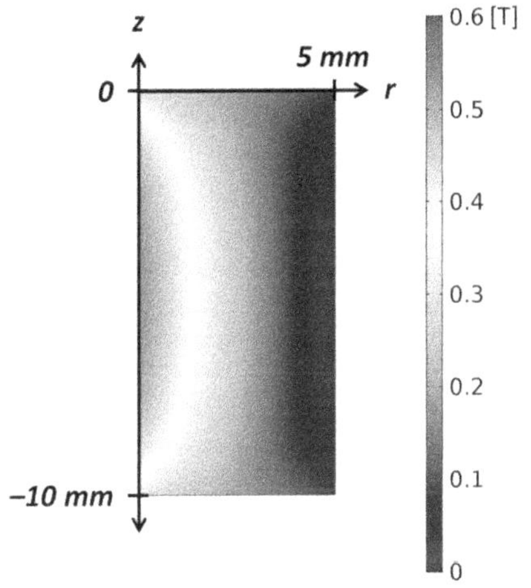

Figure 3.8. Final trapped magnetic flux density, |***B***|, throughout the cross-section of the bulk at the end of the FC magnetisation process (t = 282.5 s).

only half the external magnetising field required for the ZFC process needs to be applied and the supercurrent only flows in one direction at any time.

In the following section, two cases studies using these models are presented: (1) to model and compare MgB_2 bulks fabricated by different processes; and (2) to model iron–pnictide bulks.

COMSOL Multiphysics® software implementation:

The example model is built in COMSOL Multiphysics® version 5.4 (www.comsol.com. COMSOL AB, Stockholm, Sweden). Additional modules required: AC/DC, Heat Transfer.

(Example 2—FEM FC.mph is available at http://iopscience.iop.org/book/978-0-7503-1332-2)

If you do not have a license for the COMSOL Multiphysics® software, you can contact your COMSOL® representative at www.comsol.com/contact to request a free trial license.

3.1.1.3 Case study #1: MgB_2 bulks

In [4], a comparison is made, using a 2D axisymmetric model of FC magnetisation (see section 3.1.1.2), between bulk MgB_2 samples fabricated by the hot isostatic pressing (HIP), with and without Ti-doping, and infiltration growth (IG) methods. Table 3.1 shows detailed information on the four samples under investigation in this study. To make a fair and adequate comparison between all of the samples, which have different dimensions, the measured trapped fields are normalised using equations (2.1) and (2.2) to consider the bulk (or average), in-field J_c for each sample, averaged over the sample volume.

The measured $J_c(B, T)$ characteristics of a single, small specimen taken from each bulk sample are used as input data for the model. $J_c(B)$ at each temperature is estimated by finding suitable fitting parameters for equation (2.9) that match the measured data, and the temperature dependence of J_c, $J_c(T)$, is described using equation (2.15), where $m = 1.5$. The measured $J_c(B, T)$ characteristics and the estimated curves from the data fitting are shown in figure 3.9 and the data fitting parameters are summarised in table 3.2. J_c above/below T_c is dealt with in the same manner as shown in equations (3.1) and (3.2). In addition to the framework described in section 3.1.1.2, the resistivity is defined as

$$\rho_{\text{total}} = \frac{\rho_{\text{sc}} \cdot \rho_{\text{normal}}}{\rho_{\text{sc}} + \rho_{\text{normal}}} \tag{3.3}$$

where ρ_{sc} is from the E–J power law (equation (2.11)) as usual and $\rho_{\text{normal}} = 3 \times 10^{-8}$ Ωm, the assumed normal state resistivity of the material [5]. This enables a smooth transition from the superconducting state to the normal state when J_c is exceeded and avoids any convergence issues related to this highly nonlinear transition.

Table 3.1. Bulk MgB_2 sample information.

Sample	HIP#22	HIP#38	HIP-Ti20%	IG1
T_c [K]	38.5	38.5	39	37.5
Radius, a [mm]	11	19	18	16
Thickness, t [mm]	18	7	7	6
Aspect ratio ($2a/t$)	1.2	5.4	5.1	5.33
Relative mass density	93%	93%	94%	90%
Maximum trapped field, B_z(FC at 20 K) [T]	1.92	2.09	2.77	1.34
J_c (FC at 20 K) [A m^{-2}][a]	2.94×10^8	2.76×10^8	4.23×10^8	2.08×10^8
Reference	[5]	[5]	[5]	[6]

[a]Calculated from the measured, maximum trapped field using equations (2.1) and (2.2).

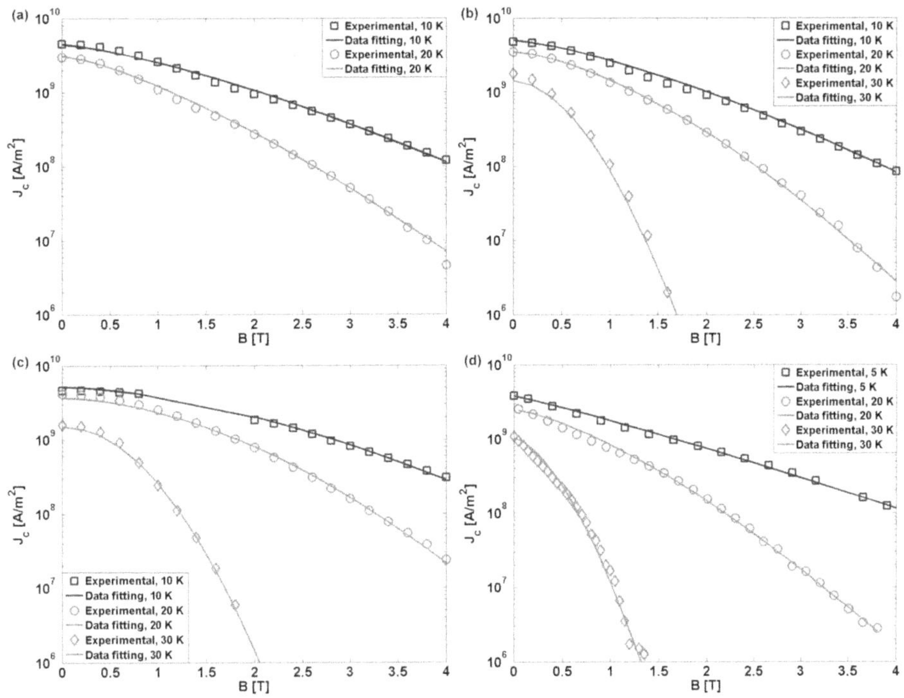

Figure 3.9. Measured $J_c(B, T)$ characteristics from a small, single specimen taken from each bulk MgB_2 sample and estimated curves from data fitting, which are used as input data for the models: (a) HIP#22, (b) HIP#38, (c) HIP-Ti20% and (d) IG1. The data fitting parameters are summarised in table 3.2.

With respect to the thermal model, the density is assumed to be 2590 kg m^{-3} and the specific heat and thermal conductivity for each sample is shown in figures 3.10 and 3.11, respectively. This data is measured from a small sample, except for the specific heat for the HIP-Ti20% sample, which is estimated using the specific heat of Ti and the standard HIP samples with a ratio of MgB_2:Ti of 0.8:0.2. This results in a slightly higher specific heat than the undoped samples (i.e., HIP#22 and #38, which show good agreement with the classical reference for MgB_2 [7]), due to the presence of Ti, which has a higher specific heat than MgB_2. Reference data for the specific heat of Ti is provided in figure 3.12 from [11]. This data is input into the thermal model using direct interpolation in COMSOL®.

It is observed that the specific heat of the IG sample is significantly higher than that of the HIP samples. After processing via the IG technique, there exists a portion of unreacted Mg and MgO in the sample (reference values are 5% MgO and 13% Mg in the sample investigated in [6]) and this unreacted Mg causes an increase in specific heat, due to the significantly higher specific heat of Mg (see figure 3.12, referenced from [13]). The specific heat of MgO is slightly lower, but similar to that of MgB_2, which is also shown for reference in figure 3.12 from [14].

Table 3.2. Data fitting parameters for the $J_c(B, T)$ characteristics of each bulk MgB_2 sample (see figure 3.9).

Sample	T_{op}	J_{c0} [A m^{-2}], equation (2.9)	B_0 [T], equation (2.9)	β, equation (2.9)	α, equation (2.15)
HIP#22	10 K	4.42×10^9	1.53	1.35	4.9×10^9
	20 K	3.1×10^9	1.05	1.35	
HIP#38	10 K	4.97×10^9	1.41	1.35	5.5×10^9
	20 K	3.48×10^9	1.09	1.51	
	30 K	1.43×10^9	0.57	1.82	
HIP-Ti20%	10 K	5.24×10^9	2.08	1.63	5.8×10^9
	20 K	3.67×10^9	1.57	1.75	
	30 K	1.51×10^9	0.75	1.98	
IG1	5 K	3.8×10^9	1.27	1.09	3.9×10^9
	20 K	2.47×10^9	0.94	1.38	
	30 K	1.02×10^9	0.39	1.59	

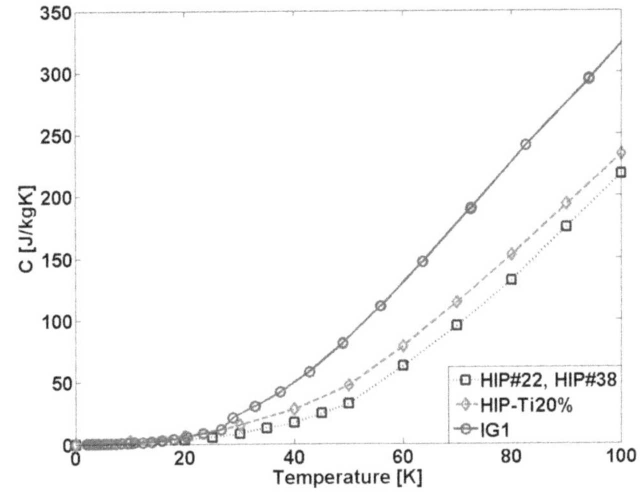

Figure 3.10. Experimentally-measured specific heat, C, data for the HIP#22, HIP#38 and IG1 samples, and the estimated C for the HIP-Ti20% sample. © IOP Publishing. Reproduced by permission of IOP Publishing from [4]. All rights reserved.

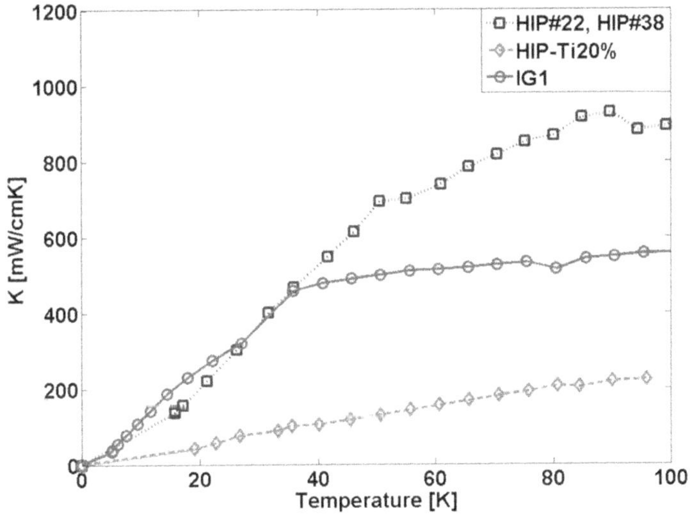

Figure 3.11. Experimentally-measured thermal conductivity, κ, data for the MgB_2 samples. © IOP Publishing. Reproduced by permission of IOP Publishing from [4]. All rights reserved.

Figure 3.13 shows a comparison of the simulation results with the temperature variation of the experimentally measured trapped field, B_z, at a height of $z = +1$ mm above the centre of the bulk surface. In the simulation, an applied field $B_{app} = 5$ T was applied to each sample to replicate the same experimental conditions under which the samples were magnetised (see, for example, [15] for the details of the FC magnetisation of the HIP samples). The time domains (see section 3.1.1.2) were split

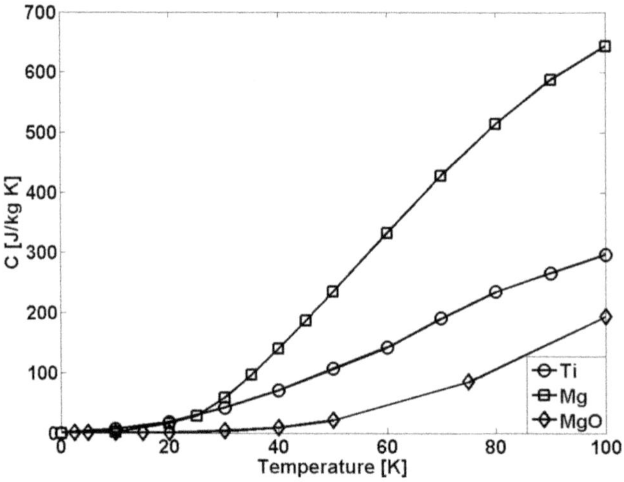

Figure 3.12. Reference specific heat values for Ti [11], Mg [13] and MgO [14].

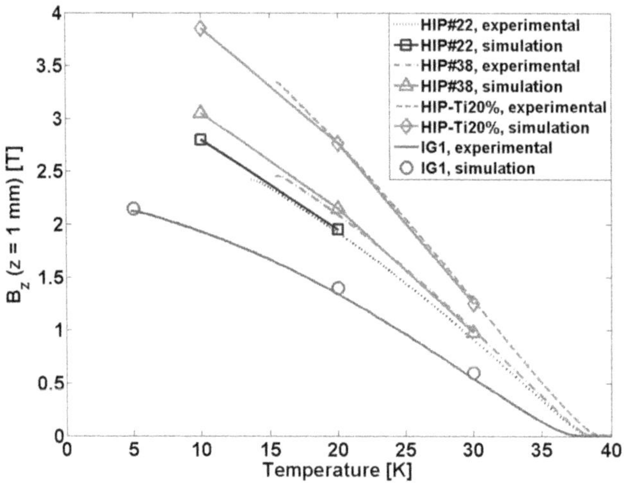

Figure 3.13. Comparison of the experimental measurements of the trapped field, B_z, with temperature and the simulation results at 10, 20 and 30 K for the HIP samples and at 5, 20 and 30 K for the IG sample.

using (1) $x_1 = 2$ s, (2) $x_2 = 60$ s and (3) $x_3 = 180$ s; this corresponds to a ramp-down rate of the applied field of 41.67 mT s^{-1}. The simulation reproduces the experimental results extremely well, and additionally, the fact that the experimental results can be reproduced with only a single, small specimen taken from each bulk sample suggests that all of the bulk samples have a highly homogeneous J_c distribution [4].

This experimentally-validated model is then used to investigate the influence of the geometric dimensions of the bulk MgB_2 sample on trapped field, based on the properties of the HIP-Ti20% and IG1 samples. It is found that, for a fixed diameter, increasing the thickness of the sample results in a slight increase in trapped field, but this saturates as the sample thickness approaches or is greater than the sample diameter. For a fixed thickness, the trapped field increases initially with increasing sample diameter, but also saturates when the diameter is several times the thickness. This is due to the drastic reduction of J_c with increasing field. Therefore, enhancing the flux pinning in order to improve the $J_c(B, T)$ characteristics of such bulk superconductors is the most effective method to increase the trapped field.

3.1.1.4 Case study #2: Iron–pnictide bulks

In [16], the magnetisation properties of iron–pnictide (Ba122) bulks are investigated using the ZFC magnetisation model of section 3.1.1.1 to simulate the FC magnetisation process by applying a large, slowly-ramped magnetising field that is several times larger than the full penetration field. Figure 3.14 shows the 2D axisymmetric models used to represent the experimental arrangement for a single Ba122 bulk, as well as a stack of multiple Ba122 bulks. A 1.6 mm spacer (not modelled) exists between the two bulks to accommodate a Hall sensor (see figure 3.14(b)).

Experimental $J_c(B)$ data, taken from a small specimen from the lower bulk, 'Ba122-lower,' measured at 4.2, 10, 20 and 25 K in fields up to 10 T, as shown in figure 3.15 (solid lines), is used. Estimated $J_c(B)$ characteristics for the upper bulk, 'Ba122-upper,' are also included in figure 3.15 (dashed lines), which are estimated from remnant trapped field data that suggested a self-field J_c, i.e., average, in-field J_c, of around 41.1 kA cm^{-2} (compared to 74 kA cm^{-2} for the lower sample). It is assumed that the upper bulk has the same field dependence as the lower one, but such that J_c is always 41.1/74 of the lower one. This $J_c(B, T)$ data is input into the model using a two-variable, direct interpolation (see section 2.2.1.3.1).

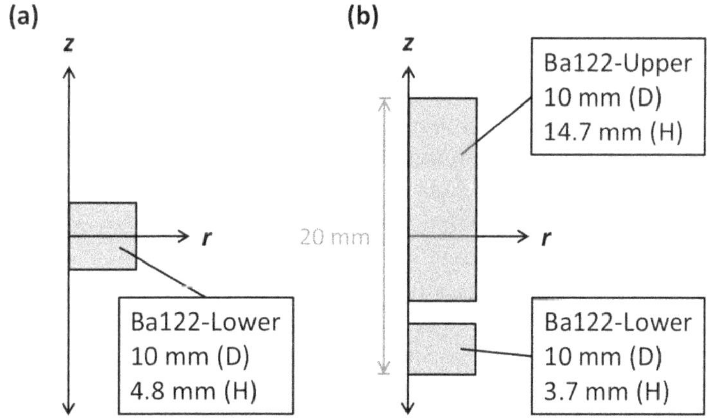

Figure 3.14. 2D axisymmetric models for numerical simulation of iron–pnictide (Ba122) bulk superconductors: (a) single Ba122 bulk, 'Ba122-lower' (diameter 10 mm, thickness 4.8 mm) and (b) stack of Ba122 bulks, comprising the same 'Ba122-Lower' bulk from (a), but polished to a thickness of 3.7 mm, and an upper bulk, 'Ba122-Upper,' comprised of bulks of diameter 10 mm and a total thickness of 14.7 mm. A 1.6 mm spacer (not modelled) exists between the two bulks to accommodate a Hall sensor. © IOP Publishing. Reproduced by permission of IOP Publishing from [16]. All rights reserved.

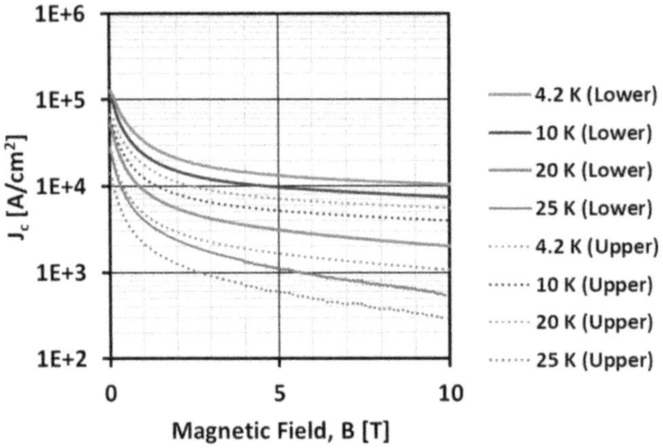

Figure 3.15. $J_c(B, T)$ characteristics, in units of A cm^{-2}, for a small specimen taken from the lower bulk, 'Ba122-lower,' measured at 4.2, 10, 20 and 25 K in fields up to 10 T (solid lines). Estimated $J_c(B)$ characteristics for the upper bulk, 'Ba122-upper,' are also included (dashed lines), which are estimated from the remnant trapped field data. © IOP Publishing. Reproduced by permission of IOP Publishing from [16]. All rights reserved.

$n = 50$ is assumed for the E–J power law, which is higher than typical values used for HTS materials, since the observed flux creep in the Ba122 bulks is comparatively much weaker. In figure 4 in [17], for example, the trapped field at 5 K decayed approximately only 3% after one day.

A uniform background magnetising field is ramped from 0 to 5 T at 1.5 T min^{-1}, then ramped back to 0 T at the same rate. After waiting +10 min for relaxation of the trapped field, the magnetic flux density is then calculated at locations corresponding to the approximate location of the Hall sensors used to take the experimental measurements. Figure 3.16 shows a comparison of the experimental and numerical simulation results for the single Ba122 bulk, 'Ba122-Lower' (see figure 3.14(a)). The trapped magnetic field, B_t, is calculated at $t = +10$ min after the application and removal of the magnetising field, at $z = +2.9$ mm, $r = 0$ mm and $z = -2.9$ mm, $r = 0$ mm, corresponding to the top and bottom Hall sensor locations, respectively (the surfaces of the bulk are located at $z = \pm 2.4$ mm). The simulation results are given for discrete points between 5 and 25 K at increments of 5 K, within the range of the input $J_c(B)$ data.

Figure 3.17 shows a comparison of the experimental and numerical simulation results for the stack of Ba122 bulks (see figure 3.14(b)). B_t is calculated again at $t = +10$ min after the application and removal of the magnetising field, at $z = -5.5$ mm, $r = 0$ mm and $z = -10.5$ mm, $r = 0$ mm, corresponding to the Hall sensor locations.

In both cases—for the single bulk and stack of bulks—the simulations reproduce the experimental results very well, validating the model assumptions, including the ZFC approximation. Since the $J_c(B, T)$ data is estimated using only a single, small specimen taken from one of the bulks, this suggests the bulks have a homogeneous J_c distribution on a macroscopic scale, in contrast to (RE)BCO bulks

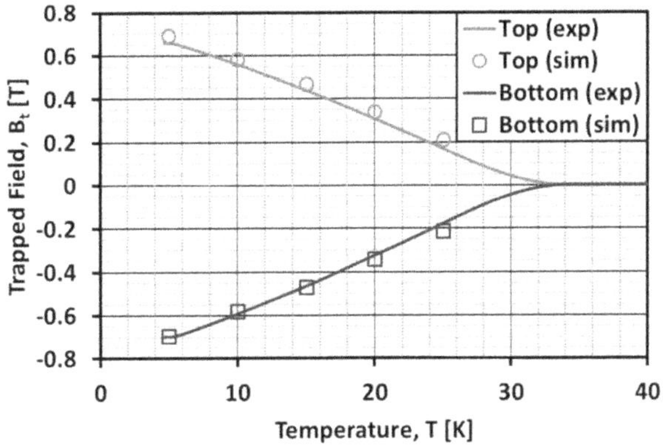

Figure 3.16. Comparison of experimental (solid lines) and numerical simulation (symbols) results for the single Ba122 bulk, 'Ba122-lower,' shown in figure 3.14(a), using the measured $J_c(B)$ data shown in figure 3.15. © IOP Publishing. Reproduced by permission of IOP Publishing from [16]. All rights reserved.

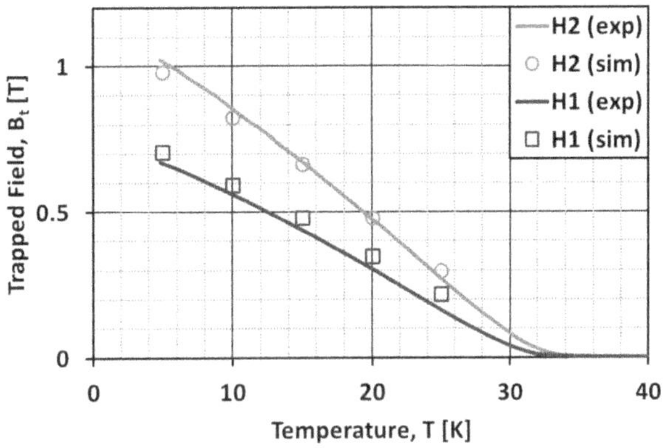

Figure 3.17. Comparison of experimental [17] (solid lines) and numerical simulation (symbols) results for the stack Ba122 of bulks shown in figure 3.14(b), using the measured and estimated $J_c(B)$ data shown in figure 3.15. © IOP Publishing. Reproduced by permission of IOP Publishing from [16]. All rights reserved.

(see section 2.2.1.3.1), although PFM is a more useful tool for highlighting the inhomogeneity of the J_c distribution [18], which is exaggerated due to local heat generation that distorts the trapped field profile.

Finally, the modelling framework is extended to investigate the influence of the geometric dimensions of the Ba122 bulk on trapped field, with a view of fabricating larger samples in the future (the samples in this study were only 10 mm in diameter). For a fixed thickness, the trapped field increases with increasing diameter, and in contrast with bulk MgB_2 materials, whose $J_c(B)$ characteristics follow an exponential law that can decrease rapidly in-field (as seen in the previous case study), rapid saturation of B_t with increasing diameter is not observed for the Ba122 bulks. There is much room for improvement to fabricate Ba122 bulks with improved $J_c(B, T)$ characteristics, but the simulation results suggests that surface trapped fields greater than 2 T could be readily achieved with state-of-the-art materials with a sample of diameter 50 mm. It also shown, by varying the thickness for a fixed diameter, that an aspect ratio of around 1–1.5 for radius : thickness would be an appropriate compromise between the accessible, surface trapped field and volume of superconducting material.

3.1.2 Pulsed field magnetisation (PFM)

3.1.2.1 Basic model

The modelling framework for PFM is essentially the same as the one for ZFC magnetisation (see section 3.1.1.1), involving the application and removal of the magnetising field to a pre-cooled bulk superconductor, except that the applied field is given by a function similar to

$$H_z(t) = H_{\max}\frac{t}{\tau}\exp\left(1 - \frac{t}{\tau}\right) \tag{3.4}$$

Equation (3.4) represents a typical, practical experimental waveform observed when discharging the stored energy in a capacitor bank through a copper magnetising coil [3, 42]. τ is an appropriate value for the rise time of the pulse, which is typically on the order of 10s of milliseconds.

In addition, the thermal model described for FC magnetisation (section 3.1.1.2) is included; however, the heat source, $Q = \boldsymbol{E}\cdot\boldsymbol{J}$, is included to account for the heat generated during the pulse, since in typical PFM experiments, τ is on the order of 10s of ms and isothermal conditions can no longer be assumed. In fact, as described in section 2.2.1.4, the electromagnetic time constant can be much smaller than the thermal time constant, especially for (RE)BCO materials. This means that the thermal diffusion can be an order of magnitude slower than the magnetic flux diffusion: the temperature rise during PFM therefore essentially takes place adiabatically.

For this example, we continue with the same bulk superconductor geometry as sections 3.1.1.1 and 3.1.1.2, the $J_c(T)$ characteristics given by equation (3.1) and an operating temperature $T_{op} = 50$ K, giving $J_c(50\text{ K}) = 3.61 \times 10^8$ A m^{-2}. Based on this J_c value, the penetration field $\mu_0 J_c a \approx 2.27$ T.

To establish the baseline peak trapped magnetic field to compare the PFM results with, firstly a long pulsed magnetic field with $\tau = 100$ s and $\mu_0 H_{max} = 4.6$ T ($\geqslant 2\mu_0 J_c a$) is applied to the bulk. In this case, the heat generated is minimal and the maximum trapped field is approximately 1.09 T at the centre of the top surface. This agrees well with a predicted value of $B_t = 1.09$ T from analytical equation (2.1).

Pulsed magnetising fields of magnitude $\mu_0 H_{max} = 4.6$ T are applied with different values of τ: 500 ms, 200 ms and 20 ms (the latter being typical of a practical experimental setup). At the end of the applied pulse ($t = 10\tau$), the calculated trapped fields are 1.04 T, 0.995 T and 0.98 T, respectively. As described in section 1.4.1, the rapid dynamic movement of magnetic flux within the superconductor during PFM results in a temperature rise ΔT, and this temperature rise increases as the rise time of the pulse is shortened, i.e., for increasing dB/dt. Figures 3.18 and 3.19 show the average and maximum temperatures, respectively, of the bulk for each pulse rise time. It is clear that the shorter pulse time results in more heat generation and a larger temperature rise, but also that for longer pulses, the electromagnetic time

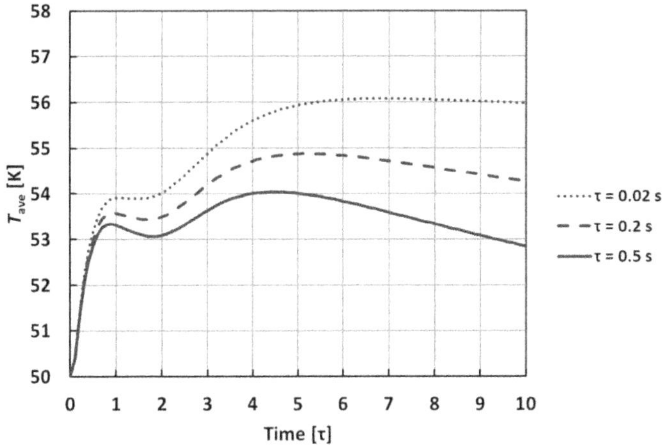

Figure 3.18. Average temperature of the bulk during the PFM process for pulsed magnetising fields of magnitude $\mu_0 H_{max} = 4.6$ T with different values for the rise time, τ: 500 ms, 200 ms and 20 ms.

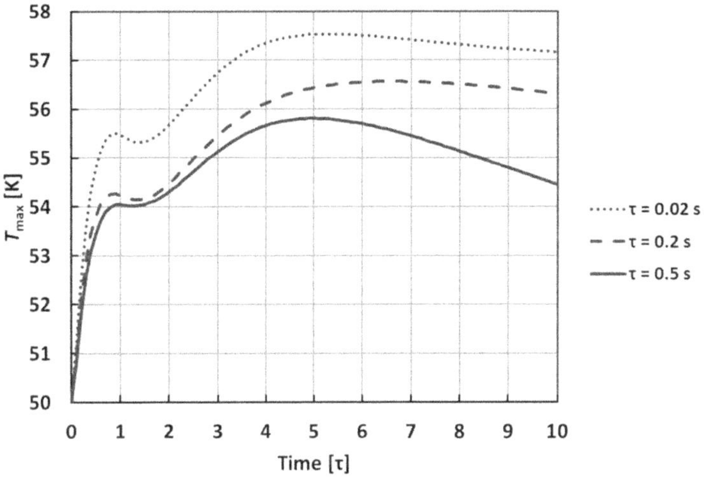

Figure 3.19. Maximum temperature of the bulk during the PFM process for pulsed magnetising fields of magnitude $\mu_0 H_{max} = 4.6$ T with different values for the rise time, τ: 500 ms, 200 ms and 20 ms.

constant is closer to that of the thermal time constant and the bulk is able to cool back down to its operating temperature more effectively.

These plots of average and maximum temperatures, however, do not give a full representation of the actual temperature distribution within the bulk and the numerical model allows us to form a detailed picture of this, which is extremely difficult to do experimentally. Figures 3.20, 3.21 and 3.22 show the temperature distribution throughout the cross-section of the bulk for rise times of 500, 200 and 20 ms, respectively, at four distinct points of time: (1) at the halfway point of the rise of the pulse, $t = 0.5\tau$, (2) at the pulse peak, $t = \tau$, (3) during the fall of the pulse at $t = 5\tau$, and (4) at the end of the pulse, $t = 10\tau$. For the shortest pulse, $\tau = 20$ ms

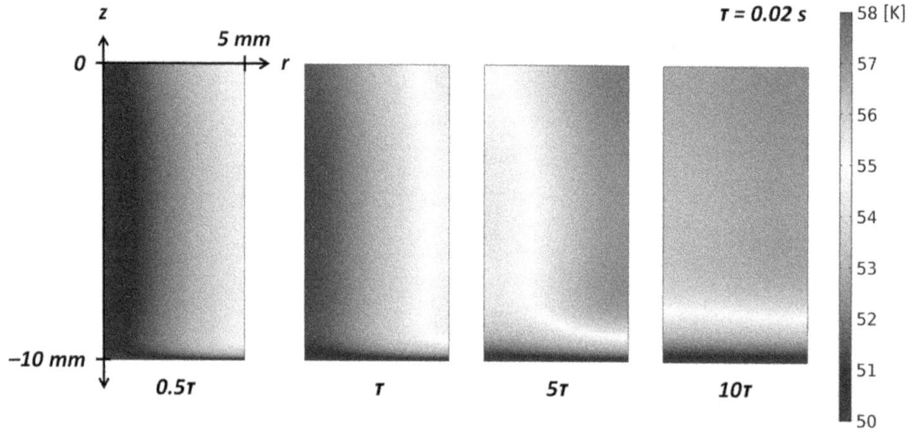

Figure 3.20. Temperature distribution throughout the cross-section of the bulk during the PFM process for a pulsed magnetising field of magnitude $\mu_0 H_{max} = 4.6$ T for a rise time, τ, of 20 ms. $t = \tau$ corresponds to the peak of the pulse and $t = 10\tau$ corresponds to the end of the pulse.

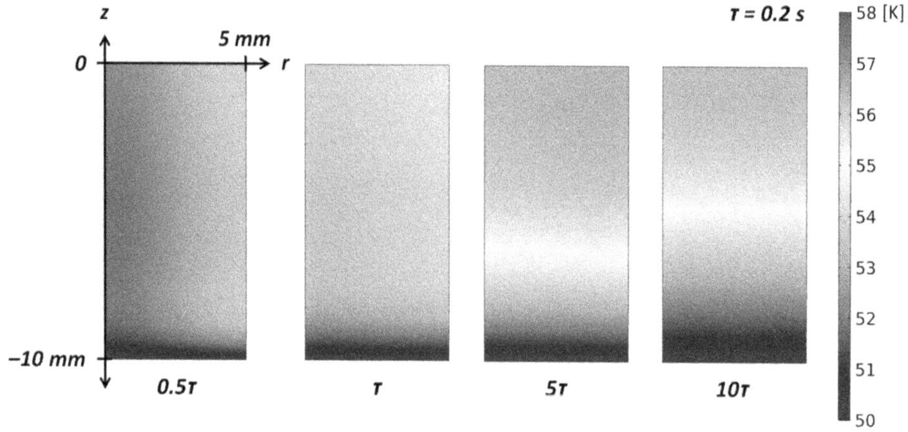

Figure 3.21. Temperature distribution throughout the cross-section of the bulk during the PFM process for a pulsed magnetising field of magnitude $\mu_0 H_{max} = 4.6$ T for a rise time, τ, of 200 ms. $t = \tau$ corresponds to the peak of the pulse and $t = 10\tau$ corresponds to the end of the pulse.

(figure 3.20), it can be seen that the process is essentially an adiabatic one, since a large amount of the heat generation takes place instantaneously and there is insufficient time for the generated heat to diffuse within the bulk and for the bulk to cool back down via the bottom surface. The temperature distribution is, therefore, quite inhomogeneous, in that heat is generated mostly at the periphery of the bulk, from where the magnetic flux penetrates into the sample. One will also notice that there is a temperature gradient between the bottom surface (connected to the cold head) and the top surface of the bulk. For the longer pulses, the temperature distribution becomes more homogeneous as the heat is able to diffuse better during the pulse. For longer pulses, less heat is generated, but also any heat generated can

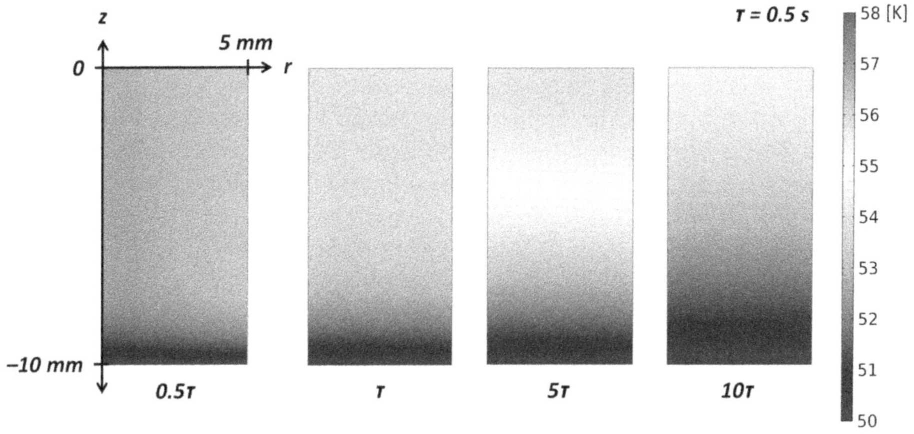

Figure 3.22. Temperature distribution throughout the cross-section of the bulk during the PFM process for a pulsed magnetising field of magnitude $\mu_0 H_{max} = 4.6$ T for a rise time, τ, of 500 ms. $t = \tau$ corresponds to the peak of the pulse and $t = 10\tau$ corresponds to the end of the pulse.

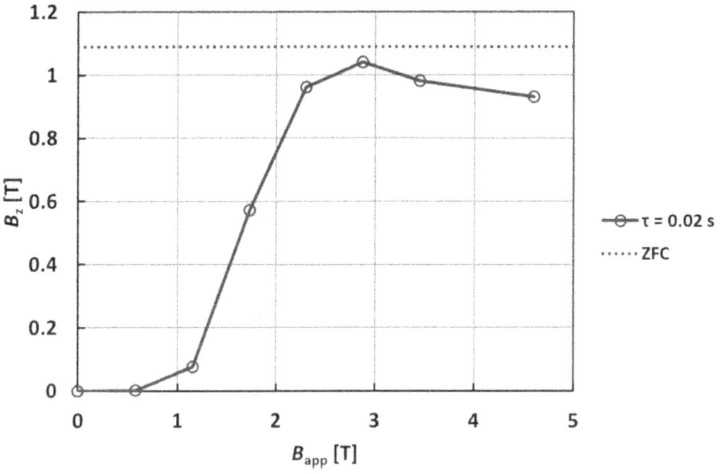

Figure 3.23. The peak, central magnetic flux density on the top surface ($z = 0$ mm) of the bulk at the end of each pulse for $\tau = 20$ ms for various applied field magnitudes, B_{app}, up to 4.6 T. For reference, the maximum trapped field capability as obtained from the ZFC model is included (1.09 T).

diffuse more easily. Eventually, for a long enough pulse, the isothermal case is reached, such as $\tau = 100$ s, which was used initially to establish a baseline peak trapped magnetic field.

Figure 3.23 shows the peak, central magnetic flux density on the top surface ($z = 0$ mm) of the bulk at the end of the pulse for $\tau = 20$ ms for various applied field magnitudes, B_{app}, up to 4.6 T. Typically, bulks magnetised by PFM are characterised in this way in order to deduce the optimal magnetising field and the corresponding maximum trapped field, recalling that the maximum trapped field capability as obtained from the ZFC model was 1.09 T. The optimal activation field is around 3 T and for applied fields larger than this, the trapped field reduces due to a larger temperature rise, although the bulk is still fully magnetised. Figure 3.24 shows the evolution of the trapped magnetic flux density profile across the top surface (z = 0 mm) of the bulk as the applied magnetising field is increased and one can observe the 'M-shaped' trapped field profile while the bulk is partially magnetised that becomes a triangular-shaped one (conical in 3D) when fully magnetised.

COMSOL Multiphysics® software implementation:

The example model is built in COMSOL Multiphysics® version 5.4 (www.comsol.com. COMSOL AB, Stockholm, Sweden). Additional modules required: AC/DC, Heat Transfer.

(Example 4—FEM PFM.mph is available at http://iopscience.iop.org/book/978-0-7503-1332-2)

If you do not have a license for the COMSOL Multiphysics® software, you can contact your COMSOL® representative at www.comsol.com/contact to request a free trial license.

In the next section, the influence of the PFM parameters on trapped fields is discussed with a particular focus on how numerical modelling has been used to explore this large parameter space.

3.1.2.2 Influence of PFM parameters on trapped fields

As mentioned in section 1.4.1, the PFM technique has many design considerations, including the magnitude and duration of the pulse(s), the number of applied pulses (e.g., multi-pulse techniques) and the type and shape of the magnetising coil/fixture (e.g., solenoid, split, vortex and so on). The temperature(s) at which the pulse(s) are applied are also crucial, as is the method of cooling, which can be done by submersion in a liquid cryogen, conduction using a liquid/gas cryogen or via a cryocooler/cold head. It is also possible to use ferromagnetic materials to shape and/or enhance the

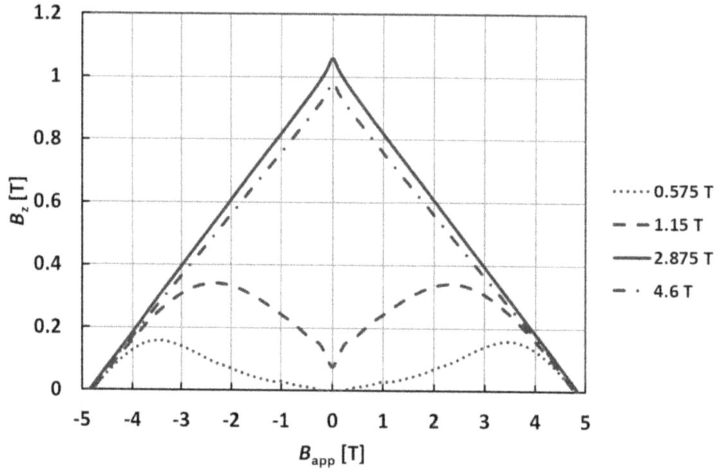

Figure 3.24. Evolution of the trapped magnetic flux density profile across the top surface (z = 0 mm) of the bulk as the applied magnetising field is increased. An 'M-shaped' trapped field profile is observed when the bulk is partially magnetised that becomes a triangular-shaped one (conical in 3D) when fully magnetised.

magnetising field and/or trapped field. Although many investigations on PFM have been carried out to date [19], there is still much of the research landscape to explore and the trapped fields achieved by PFM are still significantly lower than the true capability of bulk superconducting materials. Numerical modelling plays a crucial role here to comprehensively optimise the PFM setup in a cost-effective way, as well as propose and analyse new magnetising techniques and assess the performance of practical bulk superconductor-based applications, which are more difficult to do experimentally.

In [3], the authors use numerical simulations to investigate the effect of the pulse duration, the thermal conductivity of the bulk and cooling power of the cryocooler on the trapped field of bulk (RE)BCO. A longer pulse can enhance the trapped field because the heat generated is lower and is able to diffuse better during the pulse, as we also showed in section 3.1.2.1. However, a longer pulse demands more from the magnetising fixture: a larger capacitor bank is required, which may become prohibitively expensive, and the magnetising coil must be able to withstand large currents for a longer period of time, creating additional constraints. In section 3.1.2.1, it was assumed that the bulk was connected to a cryocooler of infinite cooling power with a perfect thermal connection; in [3], a spacing plate of finite thermal conductivity is introduced between the cold head and the bulk to a give more realistic representation of the thermal connection and cooling power of a real experimental setup. It is shown that for a better thermal connection and more cooling power, the trapped field is slightly enhanced, but the assumptions used in the model are significantly higher than values that might be achieved practically. Increasing the thermal conductivity of the bulk can also improve the trapped field: a percentage of silver can be added to the precursor powders when fabricating a bulk, which can improve the thermal conductivity by 10%–20% [20]. The reader can find more information on the thermal properties of bulk superconductors in section 2.2.1.4 and appendix A. However, this level of improvement is not enough to have a significant impact on the trapped field by PFM.

In section 2.2.1.4, we saw that bulk MgB_2 has a thermal conductivity about three times larger than bulk (RE)BCO and a specific heat per unit volume about three orders of magnitude smaller, leading to a larger temperature rise, but then with faster heat diffusion. However, combined with a narrow temperature margin between the typical operating temperature of bulk MgB_2 and its T_c, this can lead to thermomagnetic instabilities during the PFM process [21]. This manifests itself as an avalanche-like migration of magnetic flux into or out of the bulk, otherwise known as a flux jump [22]. The latter case, in particular, leads to a significant reduction in the trapped field. In [21], the authors propose safety limits for the PFM of bulk MgB_2 based on (1) an upper safety limit for the thickness of the bulk, given that subdivision of the bulk smaller than a critical thickness, d_c, can avoid a flux jump, where

$$d_c = 2\sqrt{\frac{3\gamma C(T_c - T)}{\mu_0 J_c^2}} \tag{3.5}$$

where γ is the mass density (2360 kg m^{-3} in [21]), and (2) the minimum propagation zone (MPZ) length, l_m, which is typically used to evaluate dynamic instability under adiabatic conditions, whereby the bulk can be subdivided below a size l_m and connected to a highly thermally-conductive material, then it becomes thermally stable:

$$l_m = \sqrt{\frac{2\kappa(T_c - T)}{\rho J_c^2}} \tag{3.6}$$

where ρ is the resistivity just above T_c. The authors use numerical simulations to investigate the time dependence of the local magnetic field and temperature during the PFM of bulk MgB_2 and show that the flux jump behaviour can be qualitivately understood by the local temperature exceeding T_c of the material and propose that a concentric-circled, composite bulk MgB_2 disc, consisting of many thin bulk MgB_2 rings in a highly thermally-conductive metal disc, may be a potentially valuable solution to this problem.

In [23], the authors use a model like that presented in section 3.1.2.1 to investigate these thermomagnetic instabilities in bulk MgB_2 during PFM, including realistic $J_c(B, T)$ and $n(B, T)$ assumptions and a modified E–J power law (see equation (3.3)) to explicitly include the normal state resistivity above T_c. The calculated magnetisation loops with flux jumps, for MgB_2 slabs at various temperatures, show excellent agreement with experimental results with good predictive capability. The authors also stress the necessity of including the magnetic field and temperature dependence of not only J_c, i.e., $J_c(B, T)$, but of the flux creep exponent, $n(B, T)$, too.

Flux jumps can also occur in bulk (RE)BCO and have generally been considered a hazard that lead commonly to the quench of superconducting magnets and, as much as possible, designed out in practical applications [22]. However, there has been interest recently in the possibility of exploiting flux jumps in the PFM of bulk (RE) BCO in order to minimise the shielding effect of the induced current during the rise of the pulse to both reduce significantly the magnitude of the magnetising field required for full magnetisation and increase the trapped field [12, 22, 24–27]. A theoretical framework to describe the conditions under which a flux jump would occur in type II superconductors was proposed by Wipf [28] and Swartz and Bean [29], based fundamentally on the Bean model (see section 1.4) and the assumption of an adiabatic process, which is typical of the PFM process (see sections 1.4.1 and 3.1.2.1). Assume a sample of infinite slab geometry in the superconducting state subjected to an external field, B_{app}, with a field penetration depth given by $p = B_{app}/\mu_0 J_c$. The total flux in the superconductor changes when the applied field is increased by a small amount, ΔB_{app}, by $\Delta\Phi = \frac{1}{2}\Delta B_{app} p = \frac{\Delta B_{app} \cdot B_{app}}{2\mu_0 J_c}$, which results in a dissipation of heat equal to $\Delta Q = \Delta\Phi \cdot J_c$. This reduces J_c by $\Delta J_c = \frac{\Delta Q}{C}\frac{dJ_c}{dT}$. The field shielding ability of the sample is reduced by $\Delta B_{sh} = \mu_0 \Delta J_c p$. A flux jump is then triggered if $\Delta B_{sh} \geqslant \Delta B_{app}$. The instability field for the flux jump is then given by the following equation using $\Delta B_{sh} = \Delta B_{app}$ as the critical condition;

$$B_j = \sqrt{\varepsilon \frac{\mu_0 C J_c}{| dJ_c/dt |}} \tag{3.7}$$

for which various values for the coefficient ε have been derived in the literature: $\varepsilon = \pi^2/4$ [29] based on complete local adiabaticity, $\varepsilon = 3$ [30] using the heat per unit volume averaged over the slab and $\varepsilon = 2$ using the criterion $\Delta B_{sh} = \Delta B_{app}$ [31].

The same formula can be deduced if the origin of disturbance is due to temperature rather than applied field. J_c is reduced by $\Delta J_c = \Delta T_1 \, dJ_c/dT$ when temperature is increased by ΔT_1, leading to heat dissipation and raising the temperature by ΔT_2. A flux jump will be triggered if the temperature rise is not convergent, i.e., $\Delta T_2 > \Delta T_1$.

Mints further developed these in [32] by introducing the *E*–*J* power law and the generated heat, $Q = \int \boldsymbol{E} \cdot \boldsymbol{J} \, dx$, which is closely related to the ramp rate of the magnetising field, dB/dt, from Faraday's law (equation (2.5)). Using a $J_c(B, T)$ relationship based on the Kim–Anderson relationship, $J_c(T) = \alpha(T)/B$, the flux jump field is then

$$B_j = \left[\frac{2\mu_0^2\alpha(T)h(T_c - T_0)}{n\dot{B}_{app}}\right]^{1/3} \tag{3.8}$$

where h is the heat transfer coefficient to the coolant at ambient temperature, T_0. In [22], the modelling framework in section 3.1.2.1 is used, including experimentally measured $J_c(B, T)$ of a small sample taken from a representative 15wt% Ag-containing GdBCO bulk superconductor (see figure 3.35), to predict the flux jump instability by examining when the shielding field (i.e., the magnetic field generated by the induced screening current) starts to decrease while the magnetising field is still increasing. The values of B_j obtained from the numerical model using this criterion agree well with experimental results. Flux jumps during PFM have also been reproduced qualitatively using a similar model in [12]. Such flux jumps have been reported in a number of experimental studies [22, 24, 25, 33–35] and recently a trapped magnetic field > 3 T was achieved using a portable bulk superconductor magnet system by exploiting flux jumps using multi-pulse, multi-temperature PFM [26]. Researchers have also developed a waveform control pulse magnetisation (WPCM) method by employing active feedback (magnetising coil current and magnetic flux density measured on the surface of the bulk) to control the pulse waveform using an IGBT chopper [27].

In [36], a numerical model is used to investigate the influence of bulk HTS material parameters—namely, the n value and B_0 in the Kim model—on the PFM process, using both a coupled electromagnetic-thermal model and one that ignores the thermal behaviour. The n value, as discussed in section 2.2.1.3.2, is quite difficult to determine experimentally. It is found that a smaller n value results in a larger temperature increase and stronger decay of the trapped field after the pulse. A higher B_0 value results in a higher trapped field, but also requires a larger magnetising field to fully magnetise the bulk and the heat generated during the pulse is increased. The impact on the trapped field and maximum temperature are shown in figures 3.25 and 3.26 for variation of the n and B_0 values, respectively.

In [37], a coupled electromagnetic-thermal model is used to investigate the influence of the $J_c(B, T)$ characteristics of the bulk on the PFM process using the

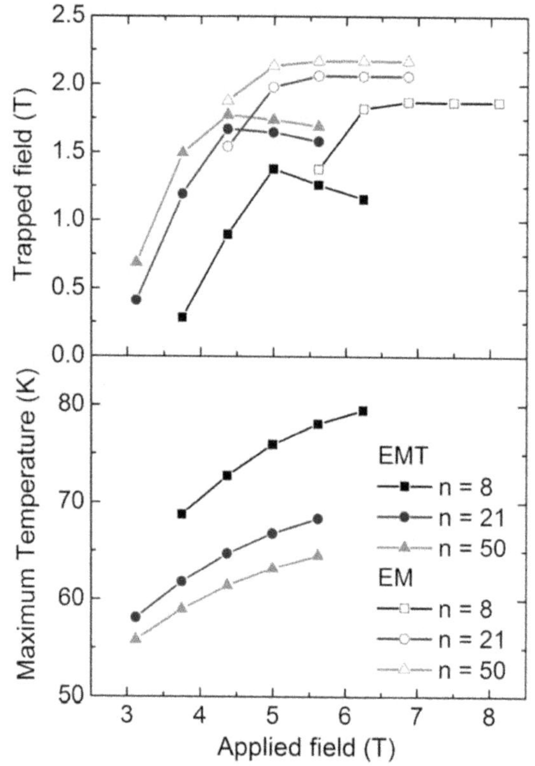

Figure 3.25. Trapped field calculated 0.8 mm above the centre of the bulk's surface and the maximum temperature with different magnitudes of applied fields after relaxation (t = 100 s). Different n values in the E–J power law are compared. EMT: electromagnetic-thermal coupled model; EM: pure electromagnetic model (T = 30 K constant). © 2016 IEEE. Reprinted, with permission, from [36].

Jirsa, Kim and Bean models (see section 2.2.1.3.1). The field-independent Bean model does not result in a higher trapped field by PFM, as one might expect, because of a larger temperature rise. The 'peak effect' in the Jirsa model results in a higher trapped field, but a steep reduction in J_c with temperature can result in a decrease in trapped field. For the Kim model, the activation field is lower than the Jirsa model due to the absence of this 'peak effect' and the trapped field is slightly larger than or similar to the Jirsa model. The authors conclude that a bulk superconductor with a J_c that has the 'peak effect' and a weak temperature dependence is optimal. The applied field dependence of the trapped field and maximum temperature rise for the three models is shown in figure 3.27 and the trapped field for the Jirsa and Kim models, with and without the thermal model, as a function of the applied pulsed field is shown in figure 3.28.

Since the trapped field from a single pulse is often a fraction of the sample's trapped field capability, a number of novel PFM techniques have been proposed. One method is the multi-pulse techniques, such as an iteratively magnetising pulsed field method with reducing amplitude (IMRA) [8] and a multi-pulse technique with

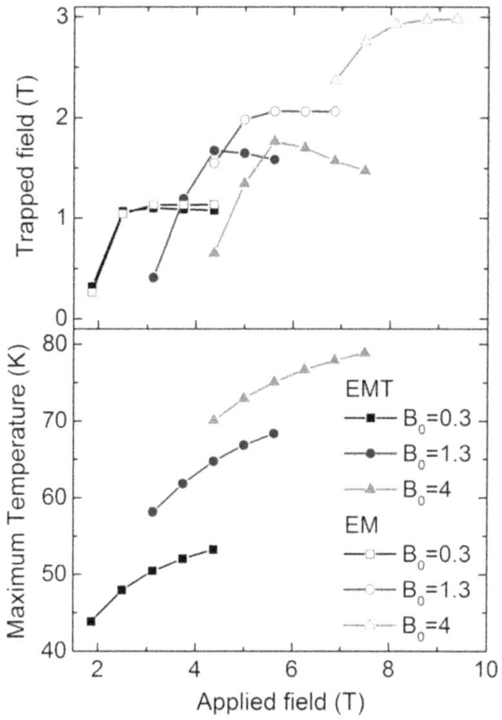

Figure 3.26. Trapped field calculated 0.8 above the centre of the bulk's surface and the maximum temperature with different magnitudes of applied fields after relaxation (t = 100 s). Different B_0 values in the Kim model are compared. EMT: electromagnetic-thermal coupled model; EM: pure electromagnetic model (T = 30 K constant). © 2016 IEEE. Reprinted, with permission, from [36].

step-wise cooling (MPSC) [9, 10]. The record achievement of 5.2 T at 29 K using the PFM was, in fact, carried out using a modified MPSC (MMPSC) technique [33, 38].

In [39], the authors use a 2D axisymmetric model, based on the critical state model and a $J_c(B)$ relation given by the Kim model, to reproduce the experimentally observed trapped field profile for a bulk that is magnetised further by a second pulsed field again after being partially magnetised by an initial pulsed field. Re-pulsing the bulk sample with appropriate temperature control is found to be effective for increasing the magnetisation for a partially magnetised sample.

In [40], the ***H***-formulation presented in [41] is extended to the numerical analysis of the MMPSC technique to compare magnetisation from a successive magnetic pulse application (SPA) method and evaluate the underlying physical mechanisms governing the process. For this qualitative analysis, the model considers an infinitely long cylinder of width the same as the diameter of the sample, an E–J power law, a linear relationship for $J_c(T)$ and adiabatic conditions for the thermal analysis. For SPA, several pulsed fields of similar magnitude are applied to a sample, which is held at a constant cooling temperature. For MMPSC, pulsed fields of relatively smaller magnitude are applied initially to a sample that is cooled to a relatively higher

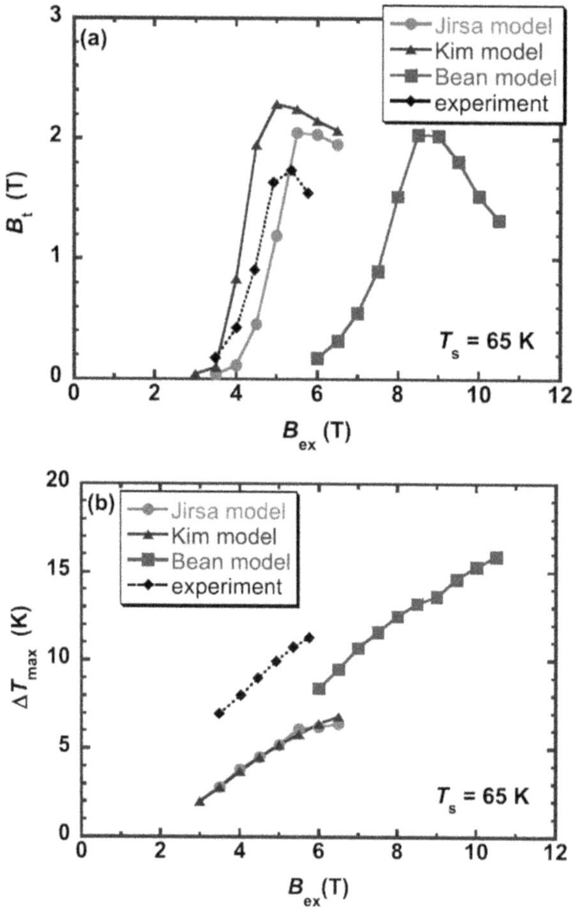

Figure 3.27. Applied field dependence of (a) trapped field, B_t, at the centre of the bulk surface and (b) maximum temperature rise, ΔT_{max}, from 65 K in the numerical simulations using three kinds of $J_c(B, T)$ characteristics. The experimental results are also shown. © 2019 IEEE. Reprinted, with permission, from [37].

temperature, and then the sample is cooled further and subjected to pulses of larger magnitude.

The numerical results for a single pulse agree qualitatively with experimental PFM results whereby increasing the magnitude of the applied pulse higher than the activation field results in an increasing reduction in trapped field [3, 42–44]. The SPA multi-pulse technique results in a trapped field that saturates with successive pulses, with no significant additional flux trapped for further pulses. The MMPSC technique, on the other hand, results in a higher trapped field after a few pulses, but a similar temperature rise in the sample in comparison to SPA. The numerical simulation confirmed the advantage of the MMPSC technique and the experimental results in [45].

In [46], the authors applied the numerical technique in [3] to evaluate the MMPSC technique. Extending the results above [40], the real cylindrical geometry

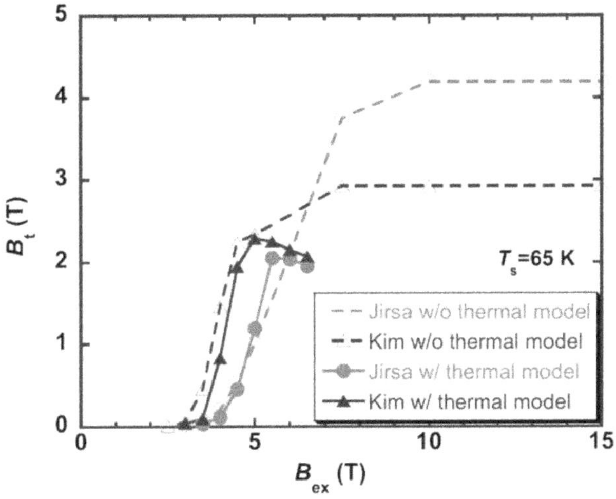

Figure 3.28. Trapped field, B_t, for the Jirsa and Kim models with (w/) and without (w/o) the thermal model, as a function of applied pulsed field, B_{ex}. © 2019 IEEE. Reprinted, with permission, from [37].

of the sample was modelled and the time evolution and position dependence of the magnetic field and temperature were reported, in order to further understand the flux dynamics and temperature propagation when using MMPSC. It was found, in particular, that the increase in trapped field after the second pulse could be explained by a reduction in temperature rise from the pulse due to the flux already trapped from the initial pulse. Similar conclusions are found in [47] using experimentally measured $J_c(B, T)$ data.

Another important practical consideration is the shape and type of the magnetising coil(s). Solenoid coils are most commonly used for fundamental PFM measurements of bulks, but a number of other coil structures have been investigated to improve the performance of the magnetising fixture and with consideration to magnetising bulks practically in TFM-based applications. Split coil magnetisation [48–52], where a pair of magnetising coils are placed above and below the bulk, instead of around the periphery of the bulk (e.g., solenoid type), has a number of merits: heat generated in the sample may be removed along the sample's *ab*-plane, which has a higher thermal conductivity than along the *c*-axis (see section 2.2.1.4); coils smaller than the diameter of the sample itself can be used, which allows for a more compact magnetisation fixture; and it has been reported that such coils can increase the total trapped flux significantly [52, 53]. A special type of split coil arrangement, known as vortex-type coils [54–58], has also been developed and implemented in an axial gap-type HTS synchronous machine [55, 59] to magnetise the rotor, which consists of bulks acting as TFMs. In [60, 61], flat racetrack-shaped magnetising coils were developed in order to use PFM to magnetise an array of bulk superconductors as field poles in a radial flux-type HTS synchronous motor. It is also possible to use superconducting materials of different T_cs and a dual cooling system to develop an *in situ* FC magnetisation process, such as the system proposed

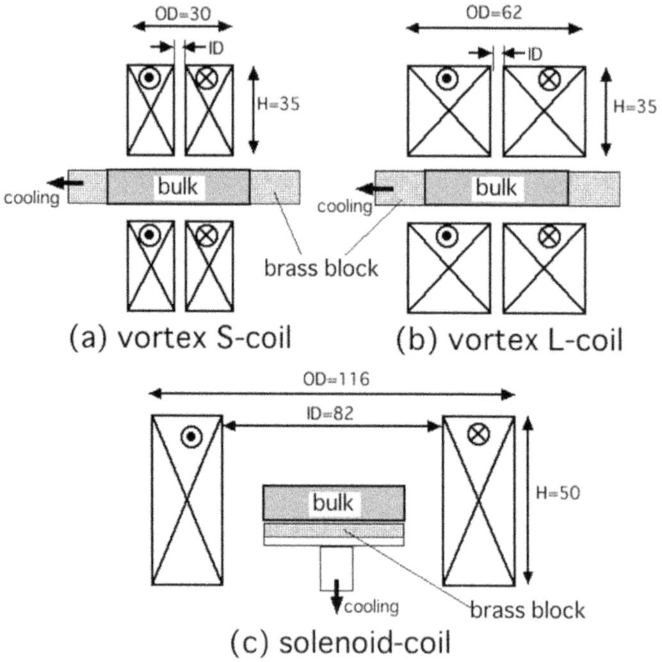

Figure 3.29. Geometric configuration for modelling vortex-type coil magnetisation of bulk superconductors: (a) vortex S-coil, (b) vortex L-coil and (c) solenoid coil.

in [62] using superconducting stator coils of an electric machine to magnetise bulk superconducting plates.

Numerical simulations of the PFM process using split- and vortex-type magnetisation coils were presented in [48] and [63], respectively. Figure 3.29 shows the geometric configuration for the reference solenoid coil and two types of vortex coils analysed in [63]: an S-coil, where the outer diameter of the coil is smaller than the bulk diameter, and an L-coil, where the coil is larger than the bulk diameter. The simulation results support many of the observed experimental results and suggest that the maximum trapped field can be larger for lower applied fields in comparison to a solenoid coil. Figure 3.30 shows the trapped field and maximum temperature rise at the centre of the bulk surface for the three coils and the results are compared with the experimental ones reported in [64]. Both [48, 63] show that with split/vortex coils, the magnetising mechanism is different in that, instead of flux penetrating from the periphery (or edge) of the bulk, the flux penetrates in from the top/bottom surfaces. The process is most effective with coils smaller in diameter than the bulk. Furthermore, it is shown that there is significantly less heat generated when pulsing with vortex coils, and a longer pulse application time, which is important in the case of solenoid-type coils [3], is not necessary for such coils. These results are promising for reducing the size of the capacitor bank required for PFM systems. Such numerical analyses play an important role to optimise the design of the magnetising

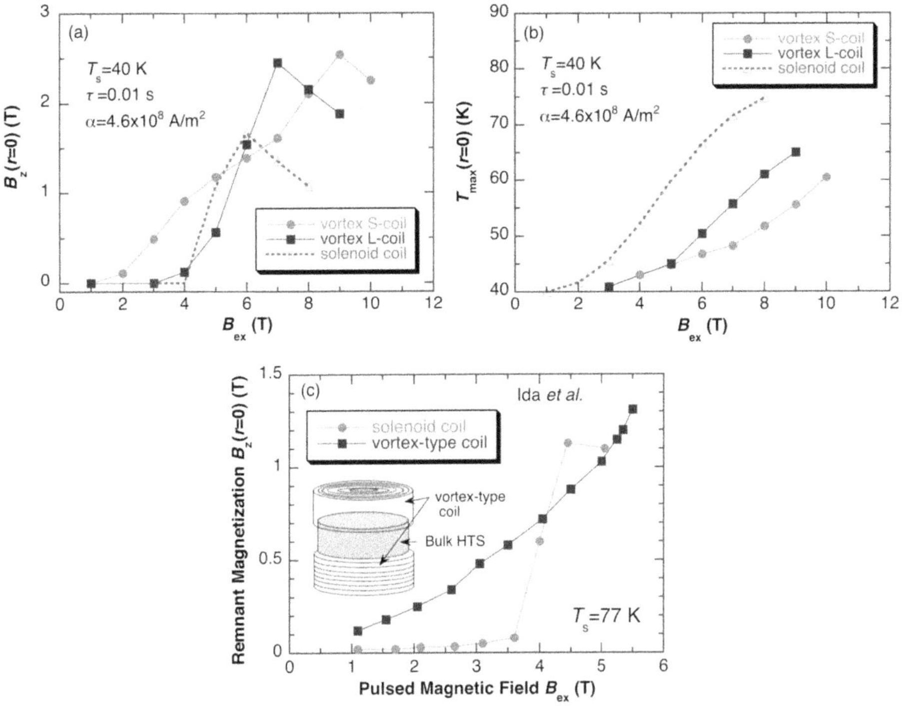

Figure 3.30. (a) Trapped field B_z ($r = 0$) and (b) maximum temperature T_{max} ($r = 0$) at the centre of the bulk surface as a function of applied field B_{ex} using the three coils (vortex S-coil, vortex L-coil and solenoid coil) shown in figure 3.29. (c) Reported experimental results by Ida *et al* in [64]. © IOP Publishing. Reproduced by permission of IOP Publishing from [63]. All rights reserved.

coil(s) to take into account practical considerations for TFM-based applications that require a compact and efficient magnetisation technique that also maximises the trapped field.

In the next section, a case study is presented, detailing how an enhanced trapped field can be achieved using a split coil PFM with an iron yoke using the numerical modelling framework in section 3.1.2.1.

3.1.2.3 Case study #3: PFM of bulk HTS materials using a split coil with an iron yoke

In [52], the authors use an extended version of the framework described in section 3.1.2.1 to qualitatively reproduce and analyse the PFM process for two types of magnetising coil: (1) a solenoid coil and (2) a split coil, both of which make use of an iron yoke to enhance the trapped magnetic field. For this study, two Ag-containing (15wt%) GdBCO bulk samples (diameter 30 mm, thickness 15 mm) were fabricated by the top-seeded melt growth process, and when magnetised by the FC process, trapped fields of $B_t = 6$ T and 5.44 T at 40 K and $B_t = 3.11$ T and 2.02 T at 65 K. Figures 3.31 and 3.32 show examples of the 2D trapped field profiles measured above the top surface ($z = +2$ mm) for one of the samples, when magnetised by (1)

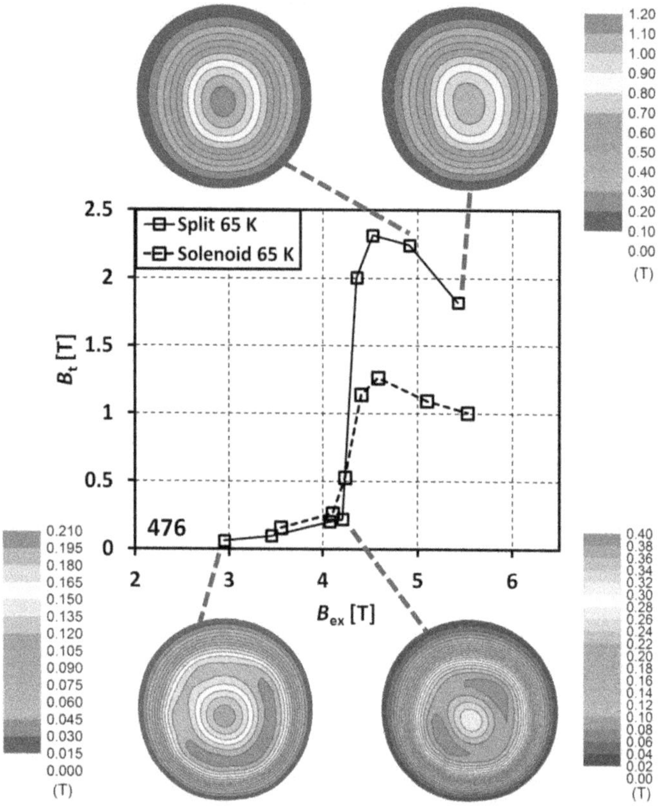

Figure 3.31. 2D trapped field profiles measured on the top surface (z = +2 mm) for one of the samples investigated in [52] after PFM using the split coil with an iron yoke at an operating temperature, T_{op} = 65 K. The central panel shows a comparison of trapped field, B_t, versus pulsed magnetising field, B_{ex}, for the two magnetising coil arrangements.

the split coil with an iron yoke (figure 3.31) and (2) the solenoid coil with an iron yoke (figure 3.32), at an operating temperature of 65 K. The 2D surface plots were taken approximately 20 min after the application of the magnetising pulse by scanning a Hall sensor using an x–y stage controller. One can observe that the magnetic flux penetrates into the sample in a fairly uniform manner, suggesting a reasonably homogeneous J_c distribution around the ab-plane. The plots of trapped field, B_t, versus pulsed magnetising field, B_{ex}, show a similar to trend to that of figure 3.23; however, one will also notice a sudden jump in B_t at $B_{ex} \approx 4.2$ T for the split coil, suggesting the occurrence of a flux jump. The split coil arrangement with an iron yoke clearly results in a significantly larger trapped field and numerical simulations are then used to verify and explain these results.

Figures 3.33 and 3.34 show the 2D axisymmetric model setups for the split coil with an iron yoke and the solenoid coil with an iron yoke, respectively. In figure 3.33, a thin (0.2 mm) indium sheet is placed around the bulk to provide a good thermal

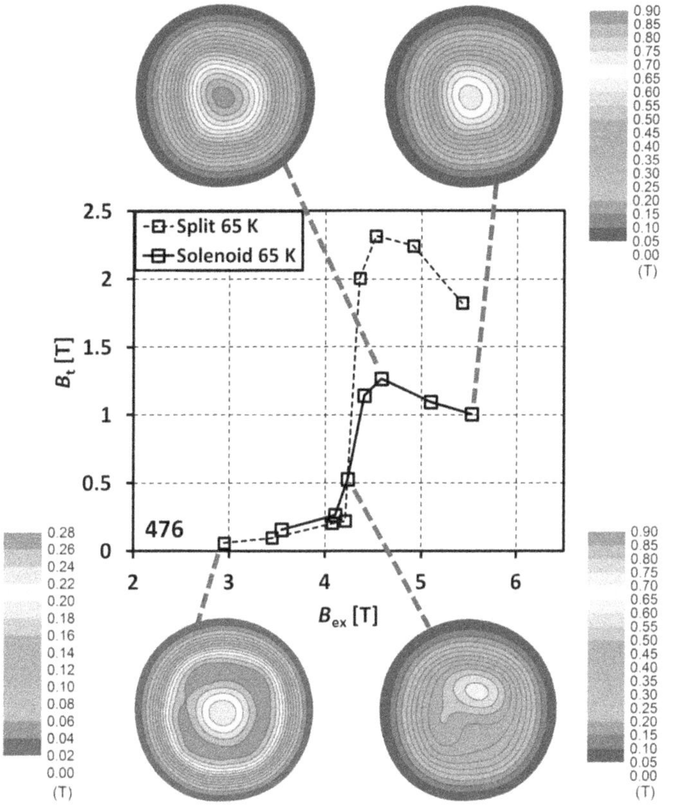

Figure 3.32. 2D trapped field profiles measured on the top surface (z = +2 mm) for one of the samples investigated in [52] after PFM using the solenoid coil with an iron yoke at an operating temperature, T_{op} = 65 K. The central panel shows a comparison of trapped field, B_t, versus pulsed magnetising field, B_{ex}, for the two magnetising coil arrangements.

contact to the sample holder. In figure 3.34, a thin (0.2 mm) indium sheet is placed between the bulk and cold stage to provide a good thermal contact and the bulk is embedded in a stainless steel (SUS) ring using Stycast™ (2850GT) epoxy.

Experimental $J_c(B, T)$ data, measured from a small sample taken from a representative 15wt% Ag-containing GdBCO bulk superconductor (diameter 30 mm, thickness 15 mm), is fit up to 10 T using the Jirsa equation (equation (2.8)) and is input into the model using a two-variable direct interpolation (see section 2.2.1.3.1). The $J_c(B, T)$ data is shown in figure 3.35 and table 3.3 shows the coefficients related to the data fitting.

The measured, temperature-dependent thermal conductivity used in the models for the SUS ring and Stycast™ (2850GT) over the temperature range 0–100 K is shown in figure 3.36, and the specific heat data for the SUS ring, Stycast™ (epoxy data from [65]), GdBCO bulk and indium sheet [66] is shown in figure 3.37. Table 3.4 summarises the other assumed model parameters, including the thermal

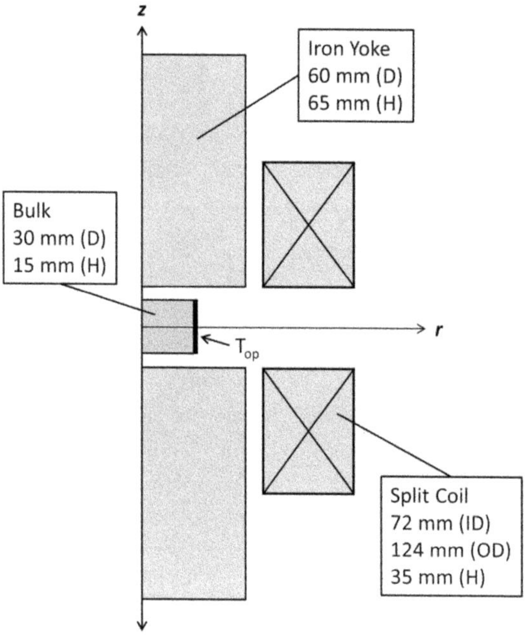

Figure 3.33. 2D axisymmetric model setup for numerical simulation of the split coil with an iron yoke. A thin (0.2 mm) indium sheet is placed around the bulk to provide a good thermal contact to the sample holder.

conductivity of the indium sheet, which is set to 0.5 W $m^{-1}{\cdot}K^{-1}$ to represent the finite cooling power of the refrigerator and the thermal contact between the bulk and the cold stage, as described in [3]. For the iron yoke, the following equation [67, 68] is used to represent the relative permeability, μ_r:

$$\begin{aligned} \mu_r &= \mu_{r,\,max} \quad \text{for } B < \mu_0 M_{sat}, \\ \mu_r &= \left(1 + \frac{M_{sat}}{H}\right) \quad \text{for } B \geqslant \mu_0 M_{sat} \end{aligned} \tag{3.9}$$

where $\mu_0 M_{sat}$ and $\mu_{r,max}$ are 1.6 T and 2000, respectively, as representative values for iron.

In order to apply the pulsed magnetising field, an integral constraint is applied to the copper coil subdomain, as described in section 2.2.1.3.3, using the following I_{pulse} function:

$$I_{pulse}(t) = N \cdot I_0 \frac{t}{\tau} \exp\left(1 - \frac{t}{\tau}\right) \tag{3.10}$$

where I_0 is the peak magnitude of the current flowing in each turn of the coil and N is the number of turns.

Figure 3.38 shows the trapped magnetic field at the centre of the top surface of the bulk sample ($r = 0$ mm) at a height of 0.1 mm at $t = 300$ ms for the split and solenoid

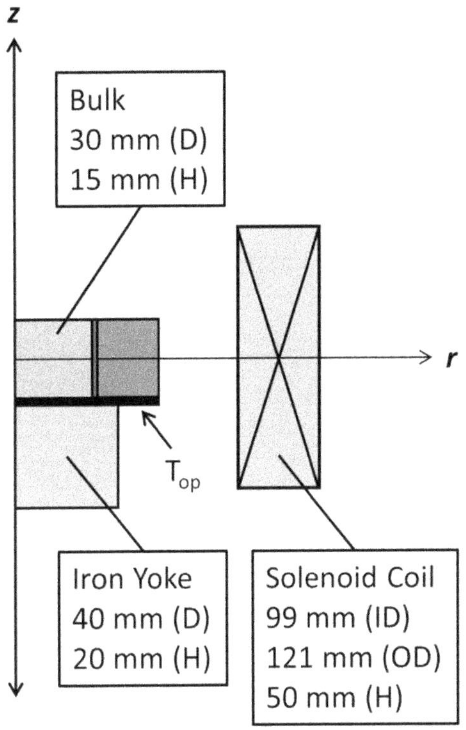

Figure 3.34. 2D axisymmetric model setup for numerical simulation of the solenoid coil with an iron yoke. A thin (0.2 mm) indium sheet is placed between the bulk and cold stage to provide a good thermal contact. The bulk is embedded in a stainless steel (SUS) ring using Stycast™ (2850GT).

coils, with and without the iron yoke, for an operating temperature, $T_{op} = 65$ K. It is clear that the presence of the iron yoke enhances the trapped field and this is particularly pronounced for the split coil, qualitatively reproducing the experimental results shown in figures 3.31 and 3.32. In the case of no yoke present, for these coil geometries, both coil arrangements result in essentially the same trapped field.

The measured trapped fields shown in figures 3.31 and 3.32 were measured before the split coil and yoke were removed for the 2D surface plot scanning measurement, and hence, the presence of the yoke would be expected to enhance the locally measured magnetic field. This results in the observed reduction in trapped field when analysing the 2D surface plots, which are also measured approximately 20 min after the application of the pulse (hence there is some flux creep; see section 1.2.3). In order to assess the 'true' trapped field, with the split coil and iron yoke removed, figure 3.39(a) shows the evolution of the trapped field from $t = 300$ ms (corresponding to the calculated trapped fields shown in figure 3.38) to $t = 10$ s, after removing the iron yoke from the split coil at $t = 4$ s. Here we examine the particular case corresponding to $T_{op} = 65$ K, $B_{ex} = 4$ T, which resulted in the maximum trapped field in figure 3.38.

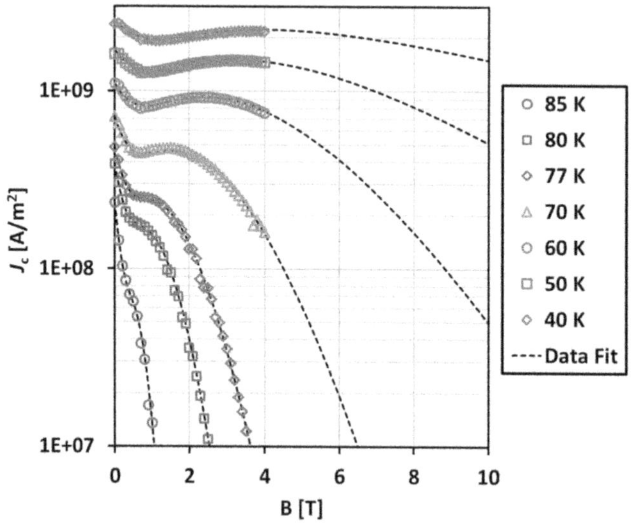

Figure 3.35. Experimental $J_c(B, T)$ data, measured from a small sample taken from a representative 15wt% Ag-containing GdBCO bulk superconductor (diameter 30 mm, thickness 15 mm). The data is fit up to 10 T using the Jirsa equation (equation (2.8)) and is input into the model using a two-variable direct interpolation (see section 2.2.1.3.1).

Table 3.3. Data fitting coefficients for the $J_c(B, T)$ characteristics of a small, representative sample using equation (2.8).

Temperature [K]	J_{c1}	B_L	J_{c2}	B_{max}	y
85	2.35e8	0.18	4.90e7	0.48	2.5
80	3.88e8	0.28	1.48e8	0.87	2
77	4.83e8	0.34	2.20e8	1.12	1.87
70	7.13e8	0.48	4.40e8	1.75	1.75
60	1.10e9	0.7	9.00e8	2.65	1.5
50	1.70e9	0.8	1.49e9	3.5	1.2
40	2.50e9	0.8	2.20e9	4.2	0.8
30	3.70e9	0.94	3.00e9	5.5	0.5

The removal of the iron yoke is simulated by reducing the $\mu_0 M_{sat}$ value exponentially to zero with a time constant of 20 ms, i.e., an exp($-t$/0.02) function, resulting in an equivalent permeability of the iron yoke of μ_0 ($\mu_r = 1$) after a period of several time constants. The dashed line in figure 3.39(a) shows the calculated trapped field when the yoke is removed at t = 4 s. There is a reduction in the observed trapped field when the yoke is removed, but the trapped field is still higher than in the case of the solenoid coil with the yoke, where the yoke remains embedded in the fixture.

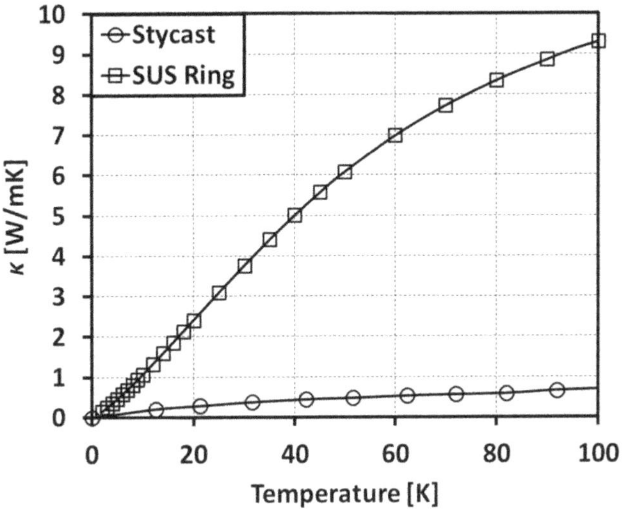

Figure 3.36. Measured, temperature-dependent thermal conductivity data for Stycast™ (2850GT) and SUS for the numerical models.

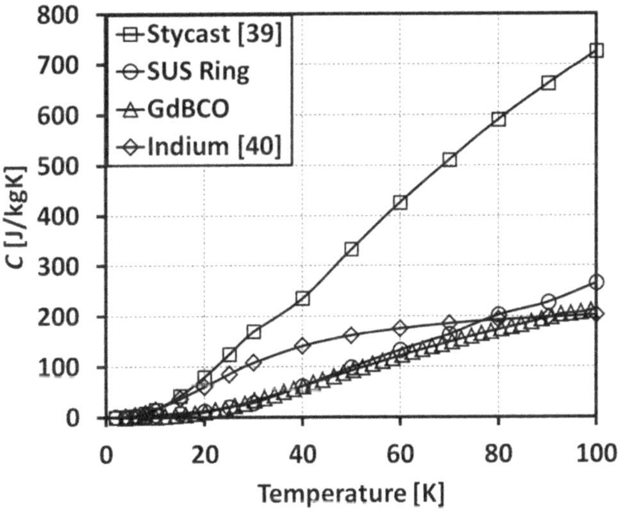

Figure 3.37. Measured, temperature-dependent specific heat data for Stycast™ (epoxy data from [65]), SUS, GdBCO bulk (15wt% Ag) and indium [66] for the numerical models.

Another important consideration is the average temperature rise during the entire magnetisation process and subsequent cooling rate for each magnetising fixture. Figure 3.39(b) shows the calculated average temperature of the bulk, T_{ave}, as the pulse increases at $t = 2$, 4, 6, 8, 10 and 15 ms (pulse peak), followed by $t = 30$, 100, 300 ms, 1, 3, 5 and 10 s. The temperature rise during the application of the pulse is very similar for both the split and solenoid coils with the yoke, but the bulk cools

Table 3.4. Other assumed model parameters. © IOP Publishing. Reproduced by permission of IOP Publishing from [52]. All rights reserved.

Parameter	Description	Value
T_c	Superconducting transition temperature	92 K
T_{op}	Operating temperature (cold head)	40, 65 K
ρ_b	HTS bulk density	5.9×10^3 kg m^{-3}
k_{ab}	Thermal conductivity of bulk along *ab*-plane	20 W m^{-1} K^{-1}
k_c	Thermal conductivity of bulk along *c*-axis	4 W m^{-1} K^{-1}
k_{indium}	Artificial thermal conductivity of indium sheet	0.5 W m^{-1} K^{-1}
E_0	Characteristic voltage in *E–J* power law	1×10^{-4} V m^{-1}
n	*n* value in *E–J* power law	21
τ	Rise time of applied pulsed current/magnetic field	15 ms

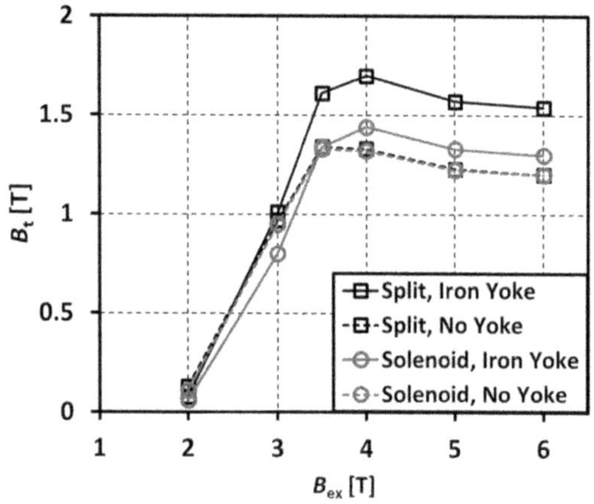

Figure 3.38. Numerical simulation results for the trapped magnetic field at the centre of the top surface of the sample ($r = 0$ mm) at a height of 0.1 mm at $t = 300$ ms for the split and solenoid coils, with and without the iron yoke, for an operating temperature, $T_{op} = 65$ K. Reproduced by permission of IOP Publishing from [52]. All rights reserved.

back down to T_{op} at a faster rate in the case of the split coil arrangement because the bulk is cooled from the periphery through the *ab*-plane, for which the thermal conductivity is several times higher (see section 2.2.1.4). Hence, the field decays slightly faster with time in the case of the solenoid coil.

Figure 3.40(a) shows the flux penetration into the sample (t = 2, 4, 6, 8, 10 and 15 ms) for $T_{op} = 65$ K, $B_{ex} = 4$ T as the pulse reaches its peak value ($t = 15$ ms). The values plotted are for B_z across the middle of the bulk from the edge to the centre. The iron yoke acts to improve flux penetration initially, but at the pulse peak, a similar flux penetration is observed, since at this magnitude of applied field, the yoke

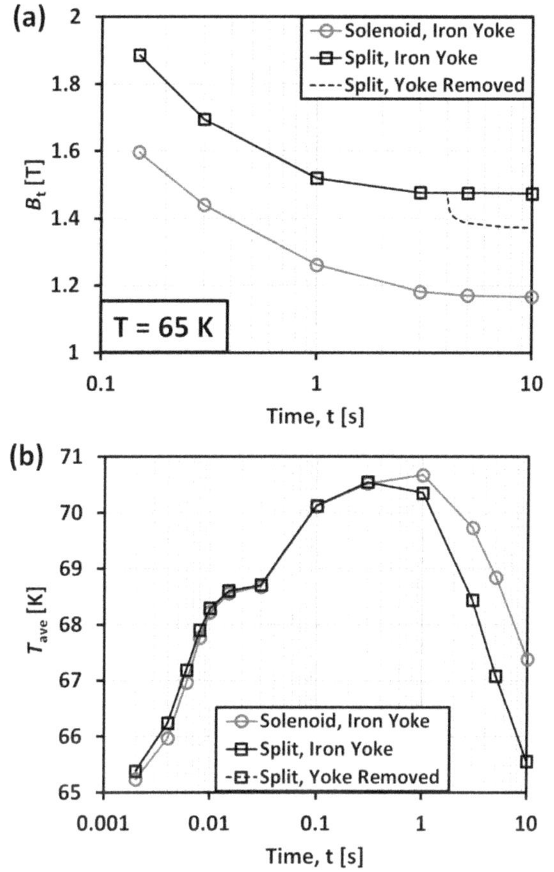

Figure 3.39. Evolution of the (a) trapped field and from t = 300 ms (corresponding to the calculated trapped fields shown in figure 3.34) to t = 10 s for the split and solenoid coil with an iron yoke, including the case where the iron yoke is removed from the split coil at t = 4 s and (b) average temperature, T_{ave}, during the PFM process. This particular case corresponds to T_{op} = 65 K, B_{ex} = 4 T, which resulted in the maximum trapped field in figure 3.38. Reproduced by permission of IOP Publishing from [52].

is saturated. As the pulse reduces and the bulk cools back down towards T_{op}, the iron yoke acts to attract flux to remain in the bulk, as shown in figure 3.40(b), for t = 30, 100, 300 ms, 1, 10 s. Significantly less flux exits the bulk when the iron core is present, resulting in a higher peak trapped field, as well as more overall trapped flux, after the magnetisation process is complete. For the solenoid coil, the enhancement of the field in comparison to no yoke can be attributed to both the difference in the flux exiting the bulk and the enhancement effect described in [67] from the presence of a ferromagnetic material on the bottom of the bulk, since the iron yoke is embedded in the magnetising fixture and cannot be removed after the magnetisation process.

The results have important implications for practical applications of bulk superconductors as such a split coil arrangement with an iron yoke could be incorporated

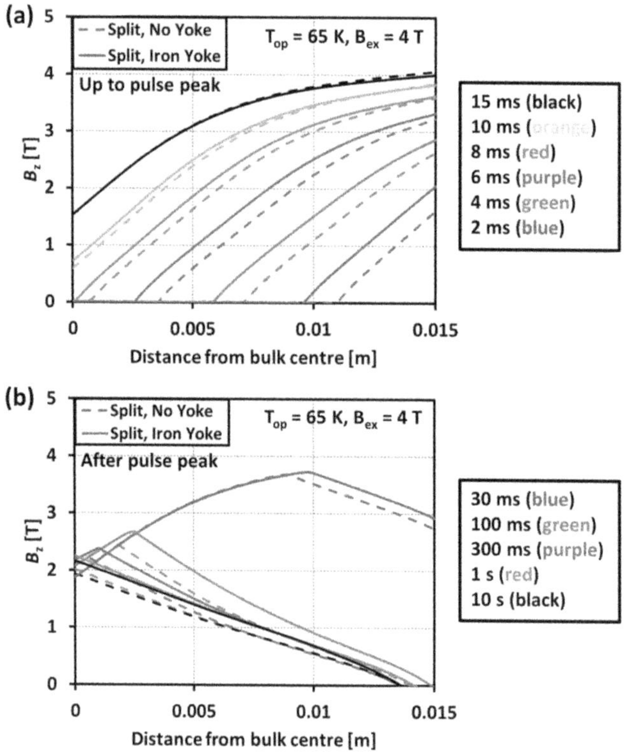

Figure 3.40. (a) Flux penetration into the sample for $T_{op} = 65$ K, $B_{ex} = 4$ T, for the split coil with and without an iron yoke, at $t = 2$ ms (blue), 4 ms (green), 6 ms (purple), 8 ms (red), 10 ms (orange) and 15 ms (black). The pulse reaches its peak value at $t = 15$ ms. (b) Flux exiting the sample as the pulse reduces and the bulk cools back down towards T_{op} at $t = 30$ ms (blue), 100 ms (green), 300 ms (purple), 1 s (red) and 10 s (black).

into the design of a portable, high magnetic field source/magnet to enhance the available magnetic field. It also has implications for an axial gap-type bulk superconducting electric machine, where iron can be incorporated into the stator windings to (1) improve the trapped field from the magnetisation process, and (2) increase the effective air-gap magnetic field.

References

[1] Ainslie M D, Weinstein R and Sawh R-P 2019 *IEEE Trans. Appl. Supercond.* **29** 6801004

[2] Sawh R-P, Weinstein R, Carpenter K, Parks D and Davey K 2017 *Supercond. Sci. Technol.* **30** 045006

[3] Fujishiro H and Naito T 2010 *Supercond. Sci. Technol.* **23** 105021

[4] Zou J *et al* 2015 *Supercond. Sci. Technol.* **28** 075009

[5] Fujishiro H, Ujiie T, Mochizuki H, Yoshida T and Naito T 2015 *IEEE Trans. Appl. Supercond.* **25** 6800104

[6] Bhagurkar A G *et al* 2015 *IEEE Trans. Appl. Supercond.* **25** 6801504

[7] Smith R M and White D 1957 *J. Am. Chem. Soc.* **79** 3641–44

[8] Yanagi Y *et al* 2000 *Adv. Supercond. XII* (Tokyo: Springer) pp 470–72

[9] Sander M, Sutter U, Koch R and Klaser M 2000 *Supercond. Sci. Technol.* **13** 841–45
[10] Kamijo H and Fujimoto H 2001 *IEEE Trans. Appl. Supercond.* **11** 1816–19
[11] Welsch G, Boyer R and Collings E W 2007 *Materials Properties Handbook: Titanium Alloys* 4th edn (Materials Park, OH: ASM International) p 143
[12] Ainslie M D *et al* 2016 *Supercond. Sci. Technol.* **29** 124004
[13] Alcock C B, Chase M W and Itkin V 1993 *J. Phys. Chem. Ref. Data* **22** 1–85
[14] Barron T H K, Berg W T and Morrison J A 1959 *Proc. R. Soc. Ser.* A **250** 70–83
[15] Yoshida T, Naito T and Fujishiro H 2014 *IEEE Trans. Appl. Supercond.* **25** 6801204
[16] Ainslie M D *et al* 2017 *Supercond. Sci. Technol.* **30** 105009
[17] Weiss J D *et al* 2015 *Supercond. Sci. Technol.* **28** 112001
[18] Fukai H *et al* 2005 *Supercond. Sci. Technol.* **18** 1179–82
[19] Ainslie M D and Fujishiro H 2015 *Supercond. Sci. Technol.* **28** 053002
[20] Fujishiro H and Kohayashi S 2002 *IEEE Trans. Appl. Supercond.* **12** 1124–27
[21] Fujishiro H, Mochizuki H, Naito T, Ainslie M D and Giunchi G 2016 *Supercond. Sci. Technol.* **29** 034006
[22] Zhou D *et al* 2018 *Supercond. Sci. Technol.* **31** 105005
[23] Xia J, Li M and Zhou Y 2017 *Supercond. Sci. Technol.* **30** 075004
[24] Weinstein R, Parks D, Sawh R-P, Carpenter K and Davey K 2015 *Appl. Phys. Lett.* **107** 152601
[25] Weinstein R, Parks D, Sawh R-P, Carpenter K and Davey K 2016 *J. Appl. Phys. Lett.* **119** 133906
[26] Zhou D *et al* 2017 *Appl. Phys. Lett.* **110** 062601
[27] Ida T, Li Z, Miki M, Watasaki M and Izumi M 2018 *IEEE Trans. Appl. Supercond.* **28** 6801905
[28] Wipf S L 1967 *Phys. Rev.* **161** 404–16
[29] Swartz P S and Bean C P 1968 *J. Appl. Phys.* **39** 4991–98
[30] Wipf S L 1991 *Cryogenics* **31** 936–48
[31] Muller K-H and Andrikidis C 1994 *Phys. Rev.* B **49** 1294–307
[32] Mints R 1996 *Phys. Rev.* B **53** 12311–7
[33] Fujishiro H, Tateiwa T, Fujiwara A, Oka T and Hayashi H 2006 *Physica* C **445–448** 334–8
[34] Yanagi Y, Itoh Y, Yoshikawa M, Oka T and Mizutani U 2005 *Supercond. Sci. Technol.* **18** 839–49
[35] Fujishiro H *et al* 2005 *Physica* C **426–431** 594–601
[36] Zou S, Zermeno V M R and Grilli F 2016 *IEEE Trans. Appl. Supercond.* **26** 4702405
[37] Hirano T *et al* 2019 *IEEE Trans. Appl. Supercond.* **29** 8000705
[38] Fujishiro H, Kaneyama M, Tateiwa T and Oka T 2006 *J. Phys. Conf. Ser.* **43** 405–8
[39] Murata K, Tsuchimoto M, Kamijo H and Fujimoto H 2004 *Physica* C **412–14** 704–7
[40] Kajikawa K *et al* 2008 *Physica* C **468** 1494–7
[41] Kajikawa K, Hayashi T, Yoshida R, Iwakuma M and Funaki K 2003 *IEEE Trans. Appl. Supercond.* **13** 3630–3
[42] Ainslie M D *et al* 2014 *Supercond. Sci. Technol.* **27** 065008
[43] Fujishiro H *et al* 2005 *Supercond. Sci. Technol.* **18** 158–65
[44] Yanagi Y, Itoh Y, Yoshikawa M, Oka T, Ikuta H and Mizutani U 2005 *Supercond. Sci. Technol.* **18** 839–49
[45] Fujishiro H, Kaneyama M, Tateiwa T and Oka T 2005 *Japan. J. Appl. Phys.* **44** L1221–4
[46] Fujishiro H, Naito T and Oyama M 2011 *Physica* C **471** 889–92

[47] Ainslie M D *et al* 2018 *IEEE Trans. Appl. Supercond.* **28** 6800207
[48] Xu Z, Lewin R, Campbell A M, Cardwell D A and Jones H 2012 *Supercond. Sci. Technol.* **25** 025016
[49] Ohsaki H, Shimosaki T and Nozawa N 2002 *Supercond. Sci. Technol.* **15** 754–58
[50] Nishijima N *et al* 2003 *IEEE Trans. Appl. Supercond.* **13** 1580–3
[51] Fujishiro H, Fujiwara A, Tateiwa T, Oka T and Hayashi H 2006 *IEEE Trans. Appl. Supercond.* **16** 1080–3
[52] Ainslie M D *et al* 2016 *Supercond. Sci. Technol.* **29** 074003
[53] Kimura Y *et al* 2006 *Supercond. Sci. Technol.* **19** S466–71
[54] Yamaguchi K *et al* 2008 *J. Phys. Conf. Ser.* **97** 012278
[55] Sano T *et al* 2008 *J. Phys. Conf. Ser.* **97** 012194
[56] Ida T *et al* 2008 *J. Phys. Conf. Ser.* **97** 012292
[57] Kimura Y *et al* 2008 *J. Phys. Conf. Ser.* **97** 012295
[58] Deng Z *et al* 2012 *Phys. Proc.* **36** 958–62
[59] Miki M *et al* 2006 *Supercond. Sci. Technol.* **19** S494–9
[60] Jiang Y, Pei R, Xian W, Hong Z and Coombs T A 2008 *Supercond. Sci. Technol.* **21** 065001
[61] Xian W, Yan Y, Yuan W, Pei R and Coombs T A 2011 *IEEE Trans. Appl. Supercond.* **21** 1171–74
[62] Masson P J, Brown G V, Soban D S and Luongo C A 2007 *Supercond. Sci. Technol.* **20** 748–56
[63] Fujishiro H, Naito T and Oyama M 2011 *Supercond. Sci. Technol.* **24** 075015
[64] Ida T *et al* 2004 *Physica* C **412–414** 638–45
[65] Iwasa Y 2009 *Case Studies in Superconducting Magnets: Design and Operational Issues* (New York: Springer) pp 354–5
[66] Jensen J E *et al* 1980 *Specific heat of some solids Brookhaven National Laboratory Selected Cryogenic Data Notebook* vol I section VIII pp VIII-C−3 (Washington DC: Brookhaven National Laboratory)
[67] Philippe M P *et al* 2015 *Supercond. Sci. Technol.* **28** 095008
[68] Zou J, Ainslie M D, Hu D and Cardwell D A 2016 *IEEE Trans. Appl. Supercond.* **26** 8200605

Mark Ainslie and Hiroyuki Fujishiro

Chapter 4

Demagnetisation and novel, hybrid superconductor structures

Chapter 4 discusses the modelling of some practical considerations for bulk superconductors, such as the effect of externally applied magnetic fields in different configurations on the attenuation of the trapped field (demagnetisation), which is crucial in applications like rotating machines where the superconductor can experience an external, time-varying magnetic field. A number of novel and hybrid bulk superconductor structures are also discussed, including composite structures with improved thermal conductivity and ferromagnet-superconductor structures, which have been designed to overcome some of the issues related to bulk superconductors and their magnetisation, and enhance the intrinsic properties of bulk superconductors acting as trapped field magnets (TFMs). Finally, the use of hollow bulk cylinders and tubes for shielding applications is discussed, as well as a new concept of a hybrid trapped field magnet lens.

4.1 Demagnetisation effects and AC losses

At finite temperatures, thermal energy can allow flux lines inside a superconductor to move from one pinning site to another due to the Lorentz force, $\boldsymbol{J} \times \boldsymbol{B}$, arising from the interaction between the current and magnetic flux density [1]. This is known as flux creep [2–4] and leads to a slow, logarithmic decay (log t) of the trapped field in a bulk superconductor over time [5] (see section 1.2.3). Hence, a consequence of the finite n value in the E–J power law relationship, given by equation (2.10), is the relaxation of the magnetisation due to flux creep, which is more suitable to describe the behaviour of HTS materials than true critical state models [6] (see section 2.2.1.3.2). Because of the logarithmic time dependence of the trapped field, the decay in trapped field is largest immediately after magnetisation, so it is important experimentally to know the time at which the trapped field measurement was taken,

doi:10.1088/978-0-7503-1332-2ch4

which can be a cause of mismatch between experimental and numerical simulation results.

When a magnetised bulk superconductor, magnetised parallel to the *c*-axis, is subjected to an external magnetic field that may be parallel or perpendicular (the so-called 'crossed-field effect' [6–9]) to its pre-magnetisation, it can suffer a partial or full decay of the trapped field [6], which is also known as the 'collapse of the magnetic moment' [8]. This can significantly reduce the bulk's potential as a TFM and can result in the failure of such bulk magnets in practical applications [10]. In a trapped flux-type superconducting synchronous machine, for example, where the HTS bulks act as TFMs (see section 1.5.2.2), (RE)BCO bulks can be employed in the rotor and magnetised parallel to the *c*-axis before the machine rotates. In these rotating machines, the position of bulk sample is expected to follow continuously the rotating field generated by the three phase stator/armature winding. In practice, however, any ripples in the applied torque on the shaft may cause the trapped field to become misaligned with the direction of stator over a short time period [11]. Due to vibration, lateral movement or an inhomogeneous stator field, bulks employed in a practical device may experience periodic variation of the external applied field [12, 13]. Although the most severe influence of external AC field arises when the field direction is perpendicular to the initial trapped field [6, 14], the trapped field can decay or even be erased by the influence of an external time-varying field parallel to the direction of trapped field even when the amplitude is much smaller than the trapped magnetic field [14–16].

In [6], the properties of bulk (RE)BCO samples were analysed numerically in the crossed-field configuration, i.e., the externally applied field is orthogonal to the magnetisation of the superconductor, using the 2D infinitely long $\boldsymbol{H}$-formulation. The numerical model qualitatively reproduced the observed experimental results, and indicated that the suppression of magnetisation comes predominately from the redistribution of the supercurrent in the bulk, leading to a rotation of the plane of symmetry of the current distribution such that it opposes the variation in magnetic flux imposed by the external field. If a field-dependent $J_c(B)$ is included in the model, then the effect is increased whereby the magnitude of J_c reduces. In [8], a similar model is used to analyse the remagnetisation process of a (RE)BCO sample partially demagnetised by an orthogonal field by applying a small magnetic field between transverse cycles that is parallel to the original magnetisation. Such models allow analysis of the dynamics of the complex supercurrent flowing in the bulk, as shown in figure 4.1, for various combinations of applied external fields [8]. Figure 4.1(a) shows the initial state of the supercurrent flowing in the full magnetised bulk, followed by the impact on this supercurrent (figures 4.1(b)–(f)) over a number of cycles due to (1) 'S-trans,' a unipolar field applied in the transverse direction (to the original magnetisation), (2) 'S-rem,' a unipolar field applied in the transverse and then parallel directions and (3) 'S-rot,' a unipolar field that is ramped initially in the transverse direction, then rotated back and forth 90° between the parallel and transverse directions.

In [17], the authors propose and experimentally validate a model to explain the rate of decay of the trapped field in a bulk GdBCO superconductor exposed to an

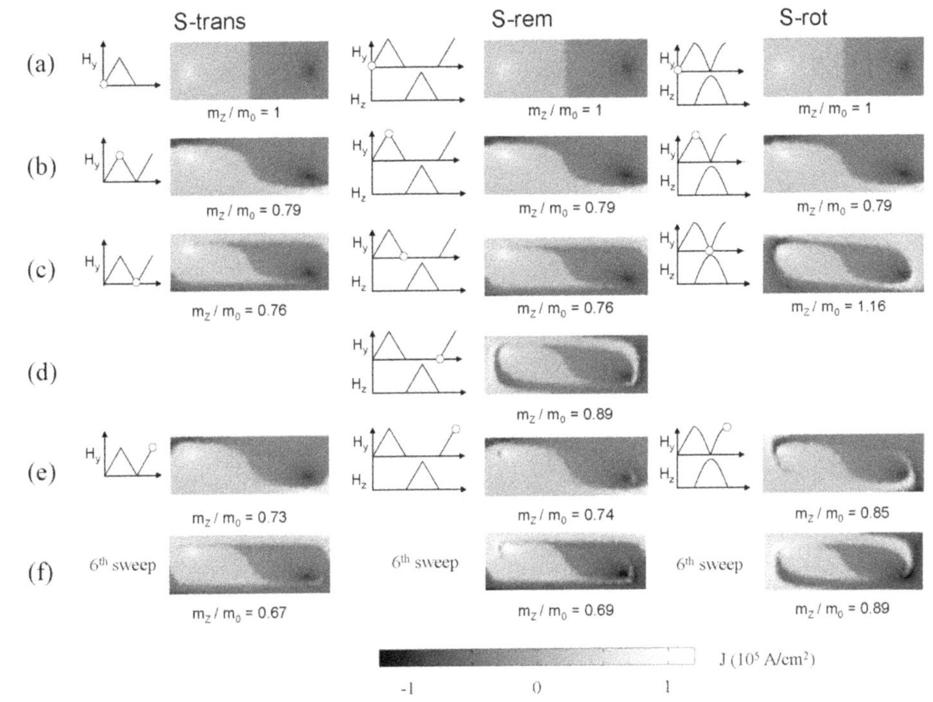

Figure 4.1. Modelled data of the complex current density distributions $J_x(y, z)$ within the cross-section of the bulk sample at selected times during each of three cycles of an applied magnetic field of magnitude $H_{\max} = 0.5H_p$. The magnetic moment normalised to the initial value, m_z/m_0, is also given below each current distribution. More details on the configuration of the applied fields can be found in the text and in [8].

AC magnetic field in the crossed-field configuration, based on calculating the time dependence of the trapped field using the Biot–Savart law and assuming that the time dependence of the current density changes at the depth of penetration of the induced shielding currents. Inside the bulk, where the crossed-field has not penetrated, the time dependence is assumed to be logarithmic and the decay of the current density due to flux creep (see section 1.2.3), whereas within the penetration depth at the surface, the time dependence is assumed to be exponential and the decay of current density is due to its redistribution. This leads to the development of two equations for the current density:

$$J = J_c\left(1 - k \ln \frac{t}{t_1}\right) \tag{4.1}$$

where J_c is the critical current density (Bean model; see section 1.4), k is the logarithmic rate of decay and t_1 is the time at which the AC field is turned on, such that $J(t = t_1) = J_c$; and

$$J = J_c \exp\left(-\frac{t - t_1}{\tau}\right) \tag{4.2}$$

where τ is a characteristic time constant of decay. Substituting equations (4.1) and (4.2) into the Biot–Savart law and integrating gives

$$B(t) = \frac{\mu_0 J_c}{2}\left\{\exp\left(-\frac{t-t_1}{\tau}\right)\left[hf\left(\frac{R}{h}\right) - (h-\lambda)f\left(\frac{R-\alpha\lambda}{h-\lambda}\right) + \lambda f\left(\frac{R-\alpha\lambda}{\lambda}\right)\right]\right.$$
$$\left. + \left(1 - k\ln\frac{t}{t_1}\right)\left[(h-\lambda)f\left(\frac{R-\alpha\lambda}{h-\lambda}\right) - \lambda f\left(\frac{R-\alpha\lambda}{\lambda}\right)\right]\right\} \quad (4.3)$$

where R and h are the radius and height of the bulk, respectively, λ is the penetration depth, α is the critical current anisotropy and $\alpha\lambda$ is thus the thickness of the shielding regions. The function, $f(x)$, is defined as $f(x) = \ln(x + \sqrt{1 + x^2})$. The terms in square brackets are constant for a given bulk geometry and dimensions, which makes equation (4.3) simply a weighted sum of an exponential and a logarithmic function. Figure 4.2 shows the shape of the time dependence, compared to that of flux creep.

Figure 4.3 shows two sets of measurements of trapped field decay on a GdBCO (10wt% Ag) bulk (diameter 30 mm, thickness 6 mm), measured at 77 K after field-cooled (FC) magnetisation in a 1.5 T electromagnet, at two different frequencies (8 and 16 Hz) and various amplitudes of applied AC magnetic field. Also shown (black lines) is the fitted analytical model, with the values of the fitting parameters, τ and k, as shown in figures 4.4(a) and (b), respectively. The authors also compare the results with a finite element model based on the 2D infinitely long ***H***-formulation, similar to the modelling framework above [8], and there is good qualitative agreement with the analytical model and experimental results.

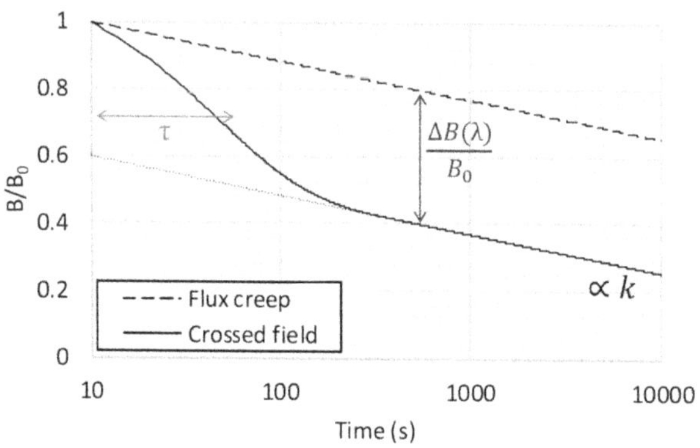

Figure 4.2. Graphical representation of equation (4.3) in a logarithmic plot (abscissa axis). The initial rate of decay is determined by the relaxation time τ, which is assigned to be 50 s in this representation. The rate of decay after several τ is determined by k, which is assumed to be 0.05. The total amount of decay due to current redistribution is determined by λ.

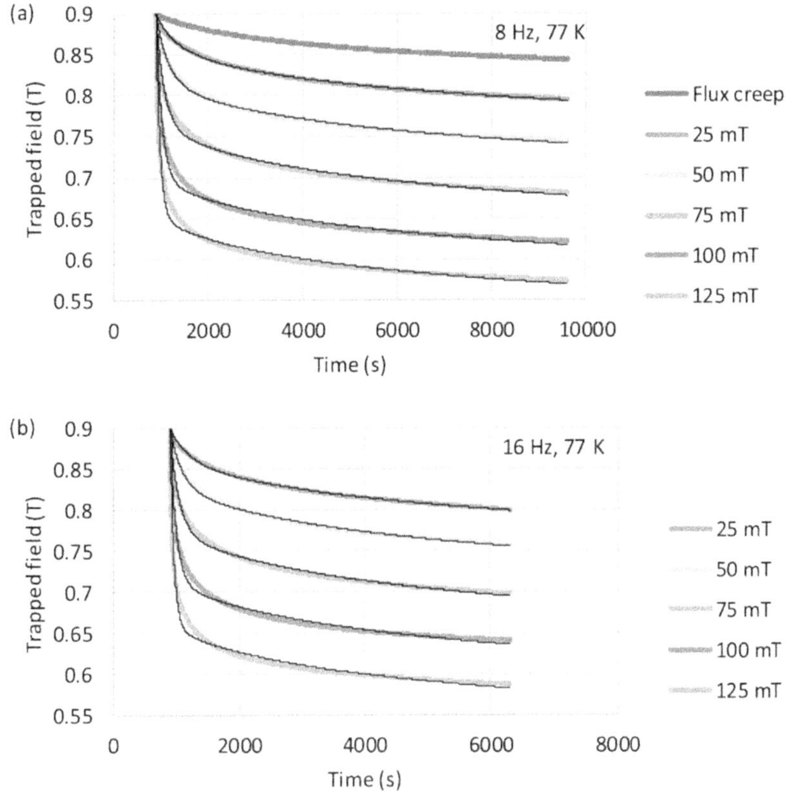

Figure 4.3. Measurement of trapped field during the application of an external AC magnetic field of various amplitudes at frequencies of (a) 8 Hz and (b) 16 Hz. The black lines represent the results of the analytical model with the fitting parameters, τ and k, shown in figure 4.4.

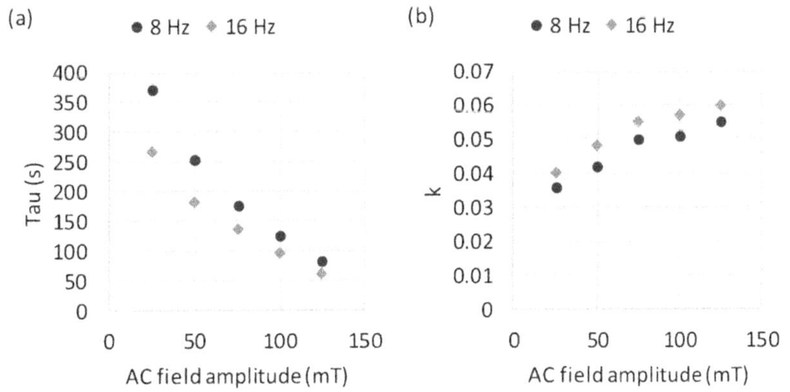

Figure 4.4. Values of the fitting parameters (a) τ and (b) k.

In [18], the authors analyse the demagnetisation of a cubic GdBCO bulk sample of equal *x*-*y*-*z* dimensions of 6 mm that is firstly subjected to FC magnetisation in a 1.3 T magnetising field, before being subjected to magnetic fields up to 164 mT in the crossed-field (transverse) configuration. The experimental results are then compared to the results from two different 3D modelling techniques: the 3D ***H***-formulation, described in section 2.2.1.2.1, and the minimum electro-magnetic entropy production (MEMEP) method, which is a variational method based on the ***T***-formulation [19].

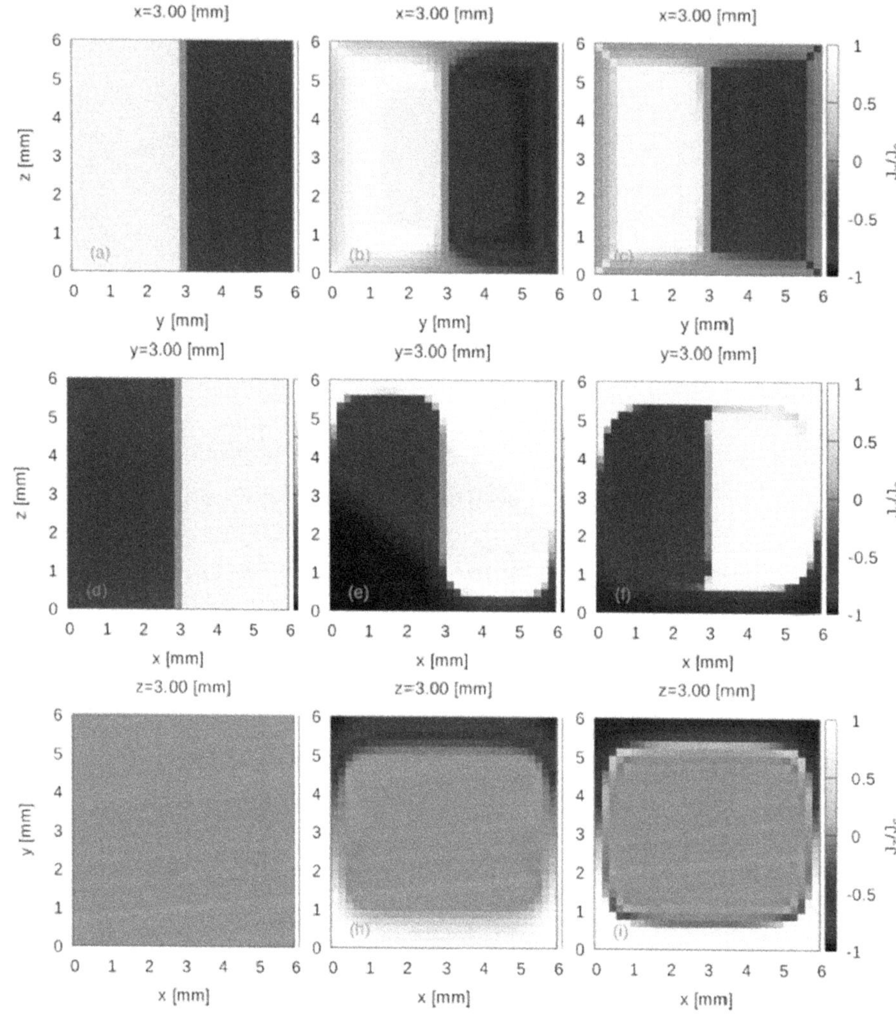

Figure 4.5. 3D penetration of the current density in the cube J_x, J_y and J_z at the centre of the cube ($x = y = z = 3$ mm) ((a), (d), (g)) for a relaxation time of 15 min after initial FC magnetisation, ((b), (e), (h)) at the first positive peak of the applied AC field (130 mT) and ((c), (f), (i)) at the last positive peak (tenth cycle of the applied field), as calculated by the MEMEP method. © 2018 IEEE. Reprinted, with permission, from [18].

Figure 4.5 shows the 3D penetration of the current density in the cube J_x, J_y and J_z at the centre of the cube ($x = y = z = 3$ mm) ((a), (d), (g)) for a relaxation time of 15 min after initial FC magnetisation, ((b), (e), (h)) at the first positive peak of the applied AC field (130 mT) and ((c), (f), (i)) at the last positive peak (tenth cycle of the applied field), as calculated by the MEMEP method. Figure 4.6 shows a comparison of the J_y component at the mid-plane at the first positive peak of the applied field for the MEMEP (left) and FEM (right) methods. Finally, figure shows a comparison of the numerical calculations (MEMEP and FEM) with the experimental measurements for AC field ripple amplitudes of (a) $B_{ax}/B_t = 1/2$, 1/4 and 1/8 for 0.1 Hz and (b) $B_{ax}/B_t = 1/2$ and 1/4 for 1 Hz, where B_{ax} is the amplitude of the AC field and B_t is

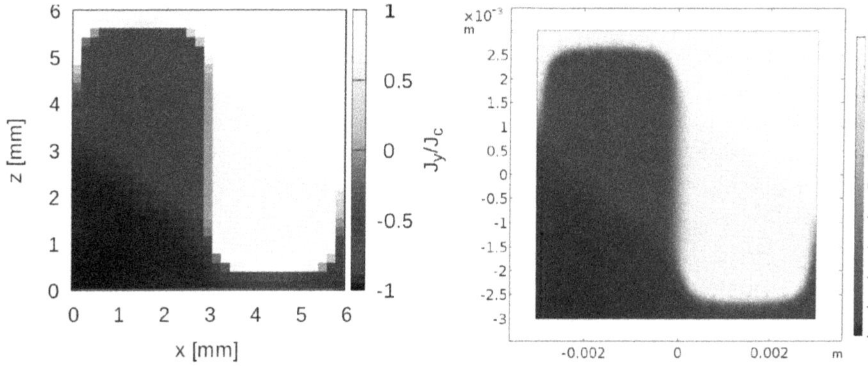

Figure 4.6. Comparison of the J_y component at the mid-plane at the first positive peak of the applied field for the MEMEP (left) and FEM (right) methods. © 2018 IEEE. Reprinted, with permission, from [18].

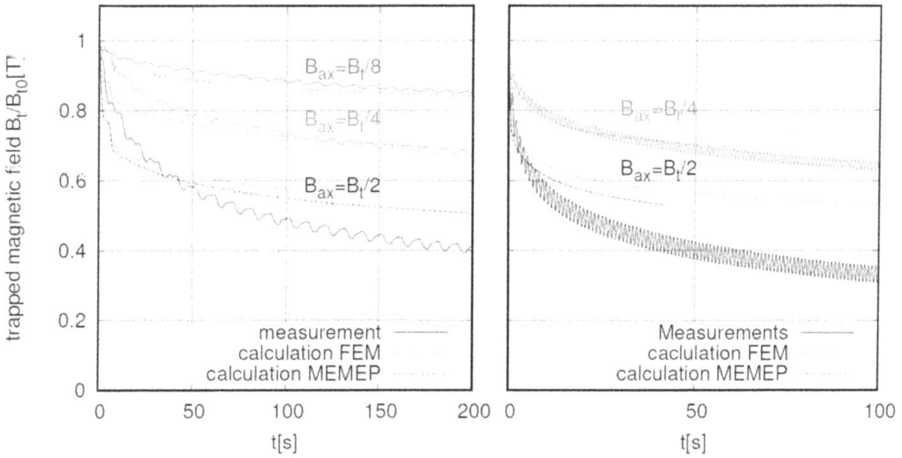

Figure 4.7. Comparison of the numerical calculations (MEMEP and FEM) with the experimental measurements for AC field ripple amplitudes of (a) $B_{ax}/B_t = 1/2$, 1/4 and 1/8 for 0.1 Hz and (b) $B_{ax}/B_t = 1/2$ and 1/4 for 1 Hz, where B_{ax} is the amplitude of the AC field and B_t is the trapped field after FC magnetisation. © 2018 IEEE. Reprinted, with permission, from [18].

the trapped field after FC magnetisation. The modelling results show good consistency and agree well with the experimental results, especially for low transverse fields, in spite of the simplified assumptions used in the models (constant, isotropic J_c and a constant n value of 30). Better agreement for higher transverse fields is expected if a more realistic n value and the magnetic field-dependence of J_c are taken into account.

In addition to the attenuation of trapped field in bulk superconductors caused by flux creep and shielding current redistribution within the bulk, consideration should also be given to changes in the thermal properties related to heat generated in the bulk. Similar to the considerations for PFM (and to a lesser degree for FC and ZFC) described in chapters 2 and 3, there is heat generated in the sample due to the penetration of an external time-varying field into the bulk, resulting in a temperature rise [20]. The experimentally observed temperature rise due to an AC field has been successfully described qualitatively using a linear temperature dependence $J_c(T)$ with Bean's model [16] or a field-dependent $J_c(B)$ and $n(B)$ [21] and assuming a constant convective heat transfer between the sample and coolant [22]. In [23], analytical expressions were derived for the steady-state (equilibrium) temperature of a bulk sample subject to an AC field based on Bean's model for an infinite slab and an infinite cylinder with a constant heat transfer coefficient. By defining two specific threshold magnetic fields, H_{tr1} and H_{tr2}, the model can predict a 'low' or 'high' steady-state temperature due to the alternating field, where thermal runaway occurs above H_{tr2}. In [24], the authors use a semi-analytical approach for a superconducting disc of finite thickness based on Brandt's algorithm [25, 26], which uses the Biot–Savart law to determine the current density $J(r, z, t)$ inside the superconductor [25–28]. This algorithm is used to compute, for one period of AC field, the power dissipated locally, $P(r, z)$, for a given $J_c(r, z)$ distribution for a constant temperature and is coupled to a heat diffusion algorithm, which is used to compute the time evolution of the local temperature, $T(r, z)$, for a given time interval Δt_H, caused by a given $P(r, z)$ distribution and specific thermal boundary conditions. The superconducting disc is discretised with steps Δr and Δz (i.e., $N_r \times N_z$ rings of width Δr and thickness Δz), as shown in figure 4.8. The time evolution of the ring temperature is calculated from the distribution of the dissipated power $P(r, z)$ using equation (4.4):

$$\rho \cdot \Delta V \cdot c_p \frac{\Delta T}{\Delta t} = \sum_{i=1}^{4} Q_i + P(r, z) \cdot \Delta V \tag{4.4}$$

where ΔT is the finite variation of temperature during the finite time interval Δt, corresponding to a number of periods of applied field (e.g., 10–20 cycles ≈ 0.2 ms). c_p, ρ and ΔV are the specific heat, density and finite volume, respectively, of each ring element, and the sum of Q_i elements are the four heat fluxes shown in figure 4.8(b): Q_{up}, Q_{down}, Q_{in} and Q_{ex}. Using such a method, the authors were able to reproduce the experimental result of heat diffusing from the corner of the sample towards the centre and investigated the space- and time-dependent temperature distribution during the application of an AC field.

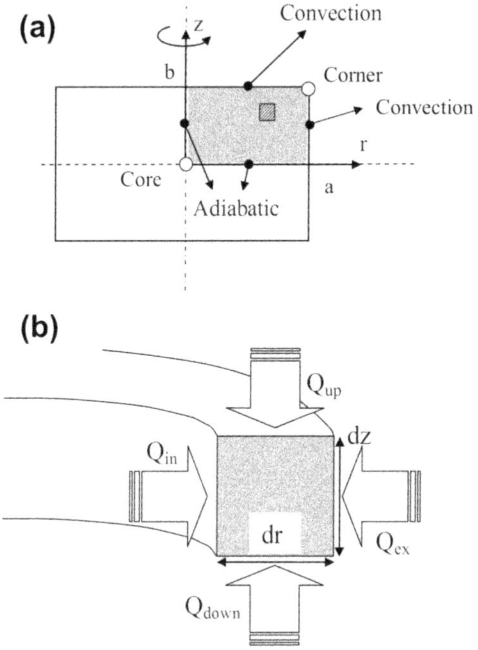

Figure 4.8. (a) Cross-section of the superconducting disc modelled in [24] showing the thermal boundary conditions. The lateral ($r = a$) and top ($z = b$) surfaces exchange heat with the cryogenic fluid by convection. The central plane ($z = 0$) and the axis ($r = 0$) are adiabatic. (b) Cross-section of a ring element of width Δr and thickness Δz. The heat flux directions, Q_{up}, Q_{down}, Q_{in} and Q_{ex}, are indicated by the arrows.

FEM techniques can also be applied to such problems and in [10] the influence of a time-varying field, applied parallel to the initial magnetisation, on trapped field attenuation was investigated using the 2D axisymmetric ***H***-formulation. Electromagnetic and combined electromagnetic-thermal models were used to investigate the influence on the trapped field profile of flux creep, shielding current redistribution and the heat generated by the penetration of the time-varying field into the bulk. Without the inclusion of thermal effects, it was found that the larger penetration depth for lower frequencies causes greater attenuation of the trapped field. However, when including the thermal model, the higher temperature rise for higher frequencies decreases the trapped field significantly. As found experimentally [14, 15], external fields of large enough magnitude can attenuate the trapped field severely, including cases where the trapped field in the bulk material is almost erased. Such models can help better explain the main causes of trapped field attenuation, with regard to penetration depth, redistribution of supercurrent, and thermal effects due to heat generated from AC losses.

Research continues to be carried out on reducing the effect of such alternating fields for practical applications using shielding [29–31], and predicting the effect of different shielding materials, such as conventional conductors, superconducting wire (which was presented in [29] using the ***A***-*V* formulation and a constant J_c approximation based on [32]) and ferromagnetic materials. The latter—hybrid ferromagnet-superconductor structures—is discussed in detail in section 4.2.2.

4.2 Novel and hybrid bulk superconductor structures

In this section, some novel and hybrid bulk superconductor structures designed to overcome a few of the aforementioned issues and enhance the intrinsic properties of bulk superconductors acting as TFMs are described, including structures with improved thermal conductivity and ferromagnet-superconductor structures. Finally, the analysis and use of hollow bulk cylinders/tubes for shielding is discussed.

4.2.1 Composite structures with improved thermal conductivity

As described in detail in section 3.1.2, heat generation can be a significant problem when carrying out PFM of bulk superconductors and can have a significant detrimental effect on the maximum trapped field from this magnetisation process. In [33], Patel and Glowacki use a numerical model to analyse the effect of various thermal conductivities in the *ab*-plane (k_{ab}) and along the *c*-axis (k_c) on the trapped field following PFM. The 2D axisymmetric ***H***-formulation is used to simulate PFM by a solenoid coil for two types of cooling environment: a thermally-isolated bulk (cooled by only minimal conduction of heat by helium gas, ignoring convection) and a bulk cooled via a cold head in a vacuum. For an isolated bulk, a higher k_{ab} increases the overall trapped flux, but decreases the peak trapped field at the centre of the top surface; k_c has a negligible effect on both. For this particular cold head system, which cools the bulk from its bottom surface increasing k_c can significantly increase the trapped field and it is unnecessary to have a high k_{ab}. However, as shown in section 3.1.2.3, the thermal dynamics of the system depend on the connection of the bulk to the cold head.

Since it is difficult to enhance the thermal conductivities of bulk (RE)BCO materials to the levels analysed in [33], the authors introduce a composite structure for the bulk using metal plates or discs embedded in the bulk, as shown in figure 4.9. It is found that structures A and C (see figure 4.9), in particular, could improve the trapped field and overall trapped flux by around 20%–30% in comparison to the bulk alone. In fact, this is considered one advantage of using stacked tape pseudo-bulks [7, 34–37], in addition to the improved mechanical strength from the presence of a metal alloy substrate and a more homogeneous and uniform J_c in the radial and axial directions for the tape [35]. Stacked tape pseudo-bulks are a relatively new area of research and numerical modelling can play an important role in the understanding and optimisation of such structures, and more concretely identify their advantages and disadvantages with respect to standard bulk superconductors.

Although bulk MgB_2 has a higher thermal conductivity in comparison to bulk (RE)BCO materials (see figure 2.5 and [33, 38]), the thermomagnetic stability of such bulks is compromised, resulting in flux jumps during magnetisation that can dramatically affect the trapped field (see section 3.1.2.2 and [39]). Indeed, such

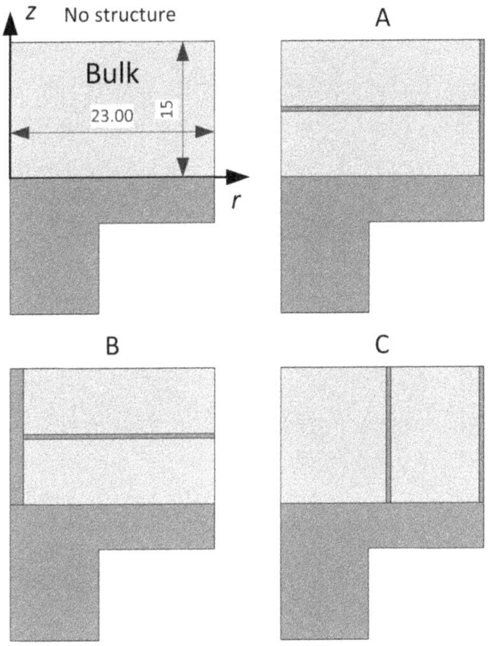

Figure 4.9. Composite bulk superconductor structures with enhanced thermal conductivity modelled in [33] for a 46 mm diameter bulk cooled by a cold head. The grey-coloured domains represents the bulk superconductor and green represents the added metal plates or discs embedded in the bulk. © IOP Publishing. Reproduced by permission of IOP Publishing from [33]. All rights reserved.

composite structures may also have a positive effect on the magnetisation of MgB_2 bulks.

The drilling of holes in the sintered powder before the crystal growth [40–42] has been proposed as a method to improve the processing of single grain bulk HTS materials, which offers a larger surface area for oxygen diffusion, enabling a higher oxygen content and a reduction in the number of macrocracks formed during oxygen annealing [43]. The mechanical strength of the TFM can be improved by impregnating the holes with a resin [42, 44] and research has also been carried out investigating the role of other filler materials, such as solder [45], copper [46] and soft ferromagnetic powder [43], for which the experimental results suggest that all of these techniques can contribute to an increased trapped field and/or flux.

In [46], a 3D electromagnetic-thermal model based on the ***H***-formulation is used to analyse the local heating effects for holes with embedded copper and it was found that heat removal is improved for such a structure. In [43], the ***A***-*V* formulation presented in [47] was used to simulate the effect of filling the holes with soft ferromagnetic powder, as shown in figure 4.10, by introducing a relative permeability μ_r for the ferromagnetic subdomains. The concentration of flux lines inside the filled holes increases the average magnetic flux density in the sample and there is an increase in trapped flux. This technique can overcome the inherent reduction in trapped field/flux resulting from drilling the holes, due to the reduced volume of

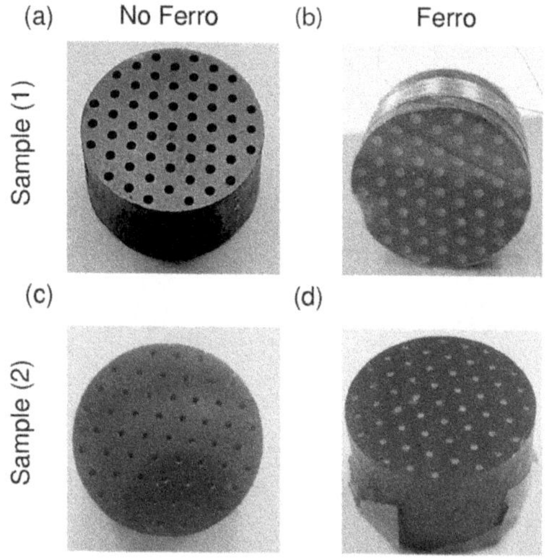

Figure 4.10. YBCO bulks with drilled holes before and after filling with soft ferromagnetic powder. (a) Sample A before impregnation. (b) Sample A after impregnation. (c) Sample B before impregnation. (d) Sample B after impregnation. © IOP Publishing. Reproduced by permission of IOP Publishing from [43]. All rights reserved.

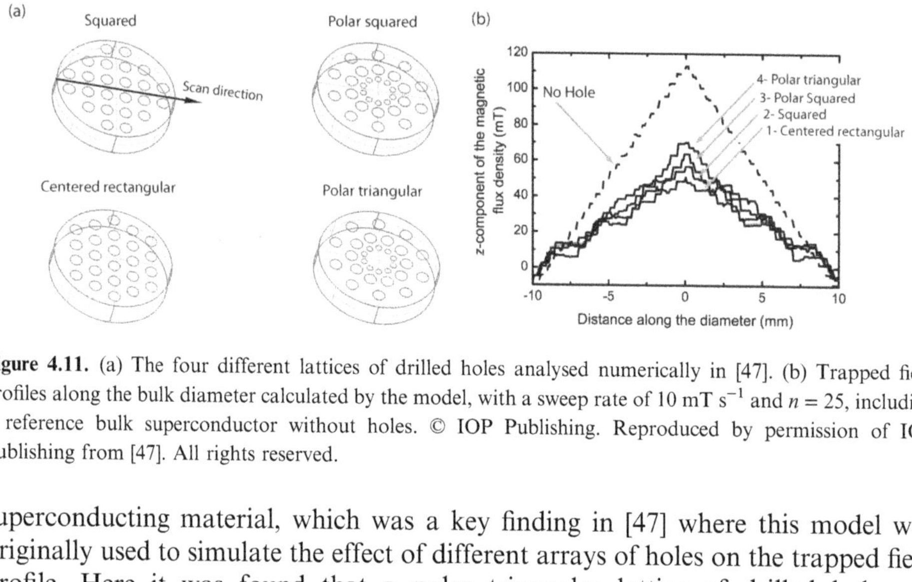

Figure 4.11. (a) The four different lattices of drilled holes analysed numerically in [47]. (b) Trapped field profiles along the bulk diameter calculated by the model, with a sweep rate of 10 mT s^{-1} and $n = 25$, including a reference bulk superconductor without holes. © IOP Publishing. Reproduced by permission of IOP Publishing from [47]. All rights reserved.

superconducting material, which was a key finding in [47] where this model was originally used to simulate the effect of different arrays of holes on the trapped field profile. Here it was found that a polar triangular lattice of drilled holes, in comparison to the other arrays shown in figure 4.11, resulted in the largest trapped field, though the trapped field is considerably less than an undrilled bulk.

4.2.2 Hybrid ferromagnet-superconductor structures

It is well known that using a combination of ferromagnetic and superconducting materials can enhance the performance of the superconductor and macroscopic hybrid ferromagnet-superconductor structures have been considered in various applications. In an attempt to reduce AC losses that occur due to the movement of magnetic flux vortices, magnetic covers, or flux diverters, have been used in a number of configurations with the wire form of superconductors: to totally enclose the filaments [48] or fill the slits between filaments [49] in a striated YBCO conductor, as a sheath around a BSCCO-2223 multifilamentary tape [50, 51] or in a C-shape to cover only the edges [52, 53] and as a horse-shoe cover for a YBCO conductor [54]. Most of these configurations resulted in a reduction in AC losses, but [50, 51] reported a loss increase, indicating that the shape and location play an important role in how effective the loss reduction is. AC loss reductions have also been reported for ferromagnetic covers on the edges of stacks of coated conductors [55, 56], and these so-called flux diverters have also been investigated for power transmission cables [57], synchronous generators [58] and transformers [59]. It is also possible to reduce the AC loss in superconducting coils using such techniques [60–63].

As described in section 4.2.1, soft ferromagnetic powder can be used to fill drilled holes in a bulk superconductor [43] and it was shown in [64, 65] that the insertion of an iron yoke in the central hole of a YBCO ring can modify the trapped magnetic field profile and enhance the trapped flux. In [66], the authors investigate the combination of a YBCO bulk with a soft FeNi ferromagnetic alloy machined into various shapes, such as cylinders and rings, as shown in figure 4.12, in order to modify and improve the trapped field profile. Numerical modelling using the Brandt method [25, 26] was used initially to determine the current distribution as a function of applied field based on equation (2.7), and Campbell's equation [67, 68] was then used to understand and compare the dynamics of the flux penetration for the hybrid structures. The soft ferromagnetic property of the FeNi alloy was incorporated by

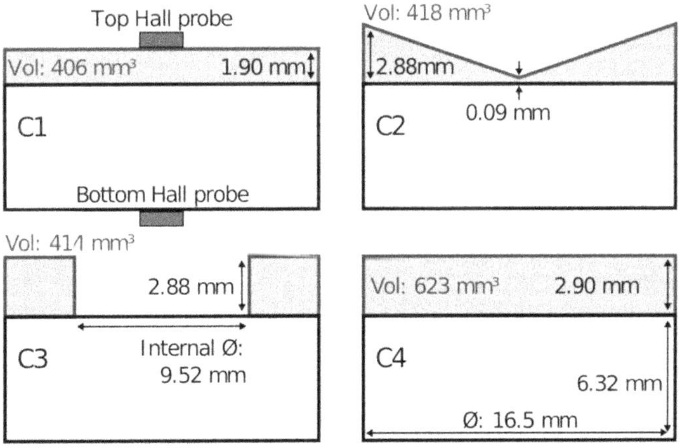

Figure 4.12. Cross-sections and dimensions of the hybrid ferromagnetic (shaded)-superconductor (white) configurations investigated experimentally and numerically in [66]. Reprinted from [66], copyright 2014, with permission from Elsevier.

modifying the left hand side of Campbell's equation from $\nabla \times (\nabla \times A)$ to $\nabla \times \left(\frac{(\nabla \times A)}{\mu_r}\right)$, where μ_r is represented by the function:

$$\mu_r = \mu_{r,\max} \quad \text{for } B_{\text{app}} = 0$$

$$\mu_r = 1 + \frac{\text{sgn}(B_{\text{app}})\mu_0 M_{\text{sat}}}{B_{\text{app}}}\left[1 - \exp\left(-\frac{\mu_{r,\max}\,|\,B_{\text{app}}\,|}{\mu_0 M_{\text{sat}}}\right)\right] \quad \text{for } B_{\text{app}} \neq 0 \tag{4.5}$$

where $\mu_{r,\max} = 1.7 \times 10^3$ and $\mu_0 M_{\text{sat}} = 1.4$ T based on experimental measurements of the FeNi alloy's *B–H* curve. A more precise fitting function can also be used to represent the field dependence of the ferromagnetic material, $\mu_r(H)$, as shown in [61, 69–71]. The numerical modelling in [66] provides a valuable insight into the difference in the contour lines of ***A*** when the ferromagnet is added to the bulk, as shown in figure 4.13, and how this affects its magnetisation. The ferromagnet acts as a shield by concentrating and deflecting the magnetic flux lines from the TFM, but the overall trapped flux is enhanced in such structures.

This work is extended in [72] for the hybrid ferromagnet-superconductor configurations shown in figure 4.14. The 2D axisymmetric ***H***-formulation presented in section 3.1.2.1 is combined with the relative permeability given by equation (3.9) in section 3.1.2.3 for Supra50 ($\mu_0 M_{\text{sat}}$ = 1.4 T and $\mu_{r,\max}$ = 1700). Figure 4.15(a) shows the experimental results for the distribution of B_z along the diameter of the bulk, 0.85 mm above the top and bottom surfaces, and figure 4.15(b) shows the modelled curves, with symmetry at r = 0 mm since the 2D axisymmetric model is used. Despite the fact that a Kim model (see section 2.2.1.3.1) approximation is used based on average magnetic measurements over the whole bulk and any position-dependence of J_c is ignored, the modelling results show excellent qualitative and quantitative agreement with the measurements. On the top surface of the bulk, there is an increase in B_z for the hybrid ferromagnet-superconductor configuration, which

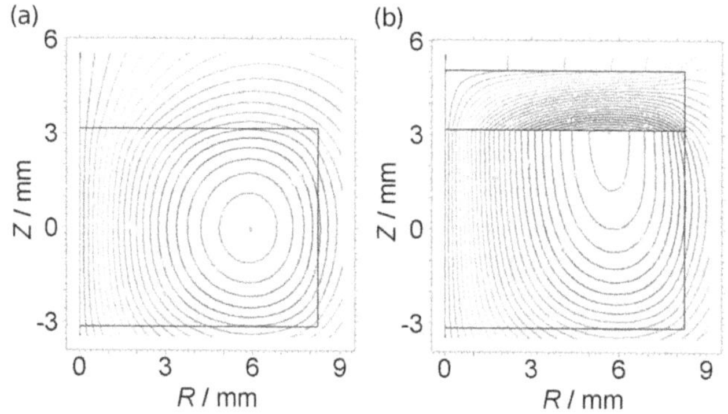

Figure 4.13. (a) Contour lines of the vector potential ***A*** in the fully magnetised remanent state for the superconductor without any ferromagnetic alloy. The results are obtained by solving Campbell's equation [67, 68]. (b) The same contour lines for the hybrid ferromagnet-superconductor structure C1 shown in figure 4.12. Reprinted from [66], copyright 2014, with permission from Elsevier.

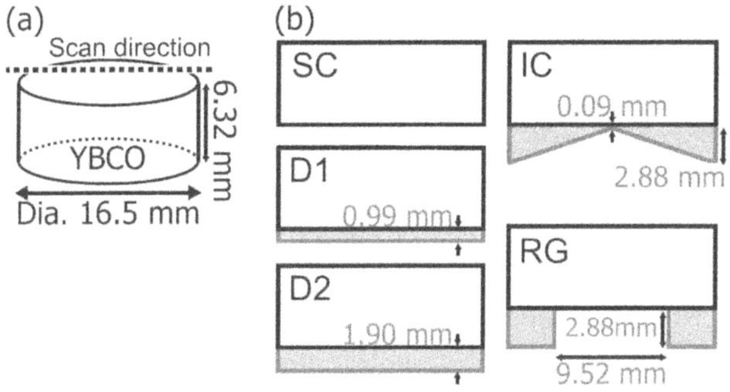

Figure 4.14. (a) Dimensions of the YBCO sample and scan direction along the diameter of the sample. (b) Cross-sections and dimensions of the measured hybrid ferromagnet-superconductor configurations. The ferromagnet (grey) is made of Supra50 and has the same external diameter as the superconductor (16.5 mm) in all the measured configurations.

depends on the shape and thickness of the ferromagnet. On the bottom surface, where the ferromagnet is located, the distribution of B_z changes from a conical one to a plate-like shape when the ferromagnet covers the entire surface and a similar profile, but with a central 'bump,' for the ferromagnet ring (configuration RG). It is also observed that a larger increase in B_z on the top surface results in a larger decrease in B_z on the bottom surface for all these configurations.

The model is then used to study new configurations, where the parameters of the hybrid configurations are changed one at a time in the model. Firstly, the influence of the ferromagnetic disc thickness is investigated for the Supra50 material. Figure 4.16 shows the modelling results for B_z 0.85 mm above and below the top and bottom surfaces, respectively, for ferromagnetic disc thicknesses varying from 0.35 mm to 2.9 mm, each with the same diameter as the bulk. B_z for the thin, 0.35 mm disc exhibits a peculiar, central 'camel-bump' on the bottom surface, which is explained using the numerical model as due to the disc being nearly fully saturated. The authors then develop an equation for the minimum ferromagnet thickness required, d^*, to avoid full saturation:

$$d^* = \frac{J_{c0}a^2}{6M_{sat}} \tag{4.6}$$

which for the actual parameters ($J_{c0} \approx 1 \times 10^8$ A m^{-2}, $\mu_0 M_{sat} = 1.4$ T and $a = 8.25$ mm) gives an estimated threshold thickness, $d^* \approx 1$ mm, which is close to that estimated by the numerical modelling (0.7–1 mm).

The influence of different saturation magnetisations for three ferromagnetic materials is then investigated: Supra50 ($\mu_0 M_{sat} = 1.4$ T as before), Permimphy ($\mu_0 M_{sat} = 0.8$ T) and Vacoflux ($\mu_0 M_{sat} = 2.2$ T). Figure 4.17 shows the modelling results for B_z 0.85 mm above and below the top and bottom surfaces, respectively, for the three materials. For higher $\mu_0 M_{sat}$ values, the general trend on the top surface is that B_z increases as a function of $\mu_0 M_{sat}$, and on the bottom surface, B_z decreases in the central region and increases near the edges of the sample, tending towards the

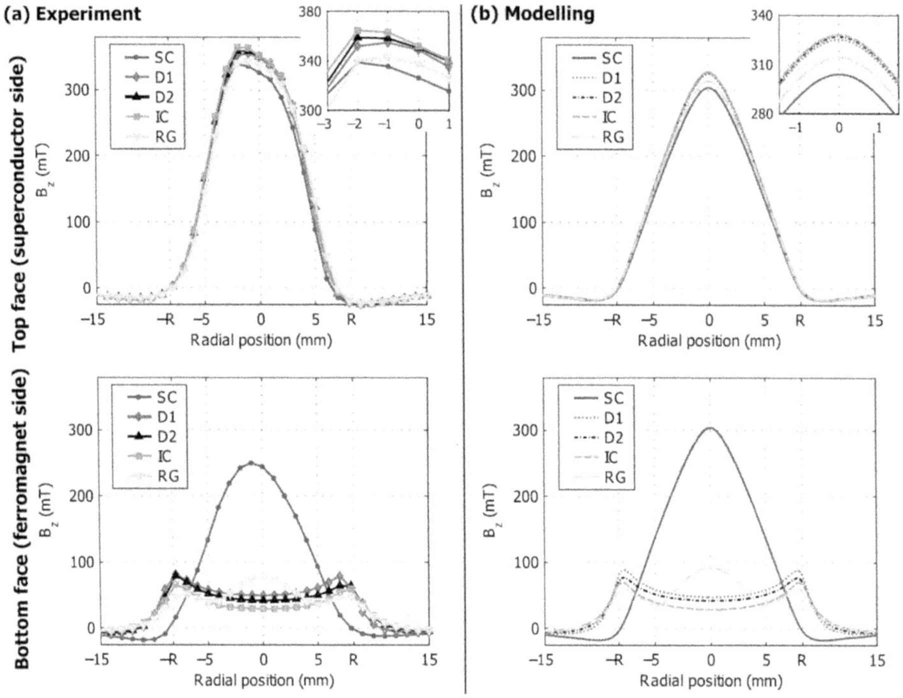

Figure 4.15. (a) Magnetic flux density B_z above the top and bottom surfaces of the hybrid ferromagnet-superconductor configurations shown in figure 4.14. All measurements are performed along a diameter 0.85 mm above the external surfaces of the whole hybrid structure. R = 8.25 mm is the radius of the ferromagnet-superconductor assembly. (b) Modelling results for the same configurations.

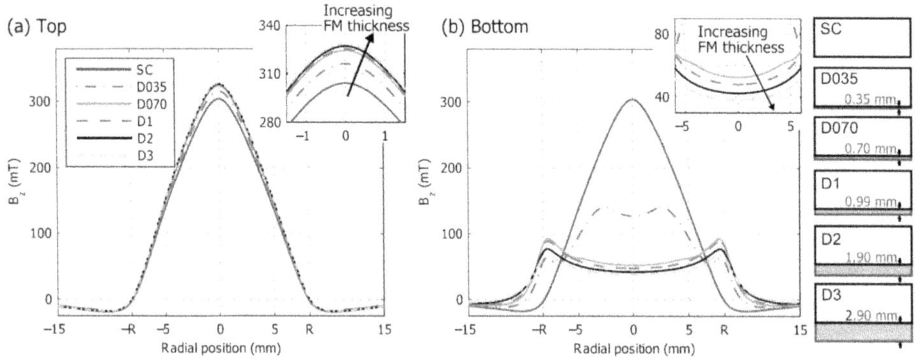

Figure 4.16. Influence of ferromagnetic disc thickness on B_z, modelling 0.85 mm (a) above and (b) under the ferromagnet-superconductor structures shown on the right hand side. The ferromagnetic discs and their thicknesses are shown in grey. The thickness of the superconductor is 6.32 mm.

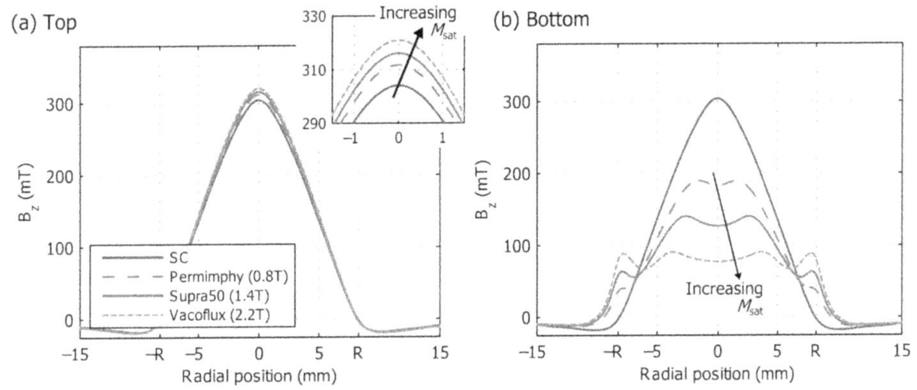

Figure 4.17. Magnetic induction B_z above the top and bottom surfaces for the D035 configuration (0.35 mm thick disc) for three different saturation magnetisations. Three materials are compared: Supra50 ($\mu_0 M_{sat}$ = 1.4 T; violet line) as presented before, Permimphy ($\mu_0 M_{sat}$ = 0.8 T; red dash), and Vacoflux ($\mu_0 M_{sat}$ = 2.2 T; green line). The superconductor alone is shown for reference (blue line).

plate-like shape observed for the thicker ferromagnetic discs (see figure 4.16). The conclusion to be drawn is that the highest $\mu_0 M_{sat}$ possible is always preferred, but the lower $\mu_0 M_{sat}$ of classical ferromagnets can be overcome by using thicker ferromagnetic discs.

In addition to enhancing and/or shaping the trapped magnetic field of a bulk superconductor, hybrid ferromagnet-superconductor structures have also been shown to be useful for reducing the crossed-field effect. In [73], the authors carry out a detailed study on the influence of the geometry of the ferromagnet on both the trapped field and crossed-field effects. Experiments are carried out on a GdBCO bulk (diameter 9 mm, thickness 5 mm) with ferromagnetic (Supra50) discs of various thicknesses attached to one or both surfaces of the bulk, as shown in figure 4.18. Both 3D (disc) and 2D (infinitely long bar) modelling of these configurations is carried out, including a model of an additional configuration: a concentric ferromagnetic ring placed around the bulk, also shown in figure 4.18. The numerical models are based on an A-ϕ formulation (3D) and an A-formulation (2D) and implemented using the open-source solver GetDP [47, 74, 75]. The superconducting part of the magnetic moment, m_{SC}, is calculated in the models by

$$m_{SC} = \frac{1}{2}\int \boldsymbol{r} \times \boldsymbol{J}\, \mathrm{d}V \tag{4.7}$$

and the ferromagnetic part of the magnetic moment, m_{FM}, is calculated by

$$m_{FM} = \int \boldsymbol{B}(\boldsymbol{r})\left(\frac{1}{\mu_0} - \frac{1}{\mu(\boldsymbol{r})}\right) \mathrm{d}V \tag{4.8}$$

The authors also introduce a correction factor when comparing the 3D and 2D models, based on the difference in the ratio between the ferromagnetic and the

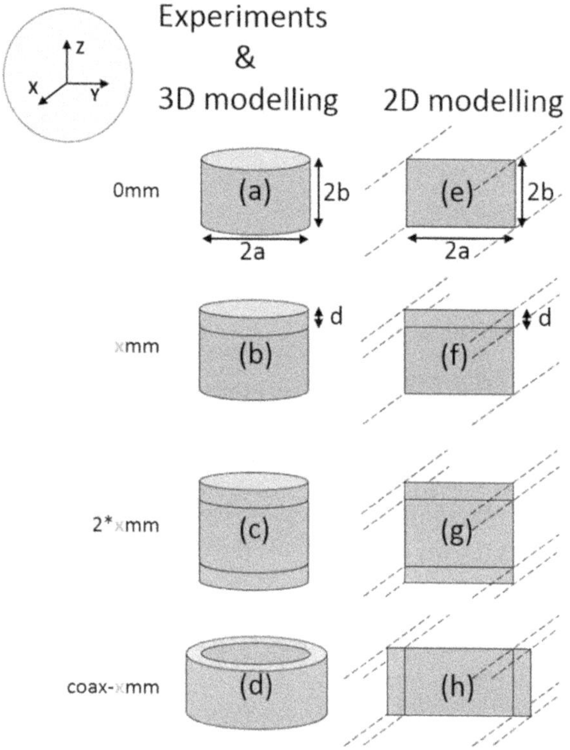

Figure 4.18. Schematic representations of the studied geometries in experiments (a)–(c), 3D modelling (a)–(d) and 2D modelling (e)–(h). The bulk superconducting pellet/bar (diameter/width $2a$ and height $2b$) is labelled 0 mm ((a) and (e)). The hybrid structures consisting of the same bulk pellet/bar and ferromagnetic discs/bar of several thicknesses ($d = x$ mm) attached to the top surface are labelled 'x mm' ((b) and (f)). The hybrid structure where two ferromagnetic discs/bars are attached to the top and the bottom surfaces is labelled '2*x mm' (c) and (g). In 3D, the hybrid structure where a ferromagnetic ring is placed around the bulk pellet is labelled 'coax-x mm' (d), as is the case in 2D, where two ferromagnetic bars are attached to the sides of the bulk bar (h).

superconducting magnetic moments, such that the 2D models are corrected by a factor ¾ (see [73] for more details).

Figure 4.19 shows a comparison of the experimental measurements of the magnetic moment and the numerical results for the 3D and 2D models, normalised to the value when no ferromagnet is present (i.e., only the bulk). The 3D model shows excellent agreement with the experimental results, but the 2D model over-estimates the magnetic moment; after correcting the 2D modelling results as discussed above, there is much better agreement.

The authors then investigate the penetration of the shielding currents from a crossed-field for two situations (with and without a 2 mm ferromagnetic layer on the top surface of the bulk): (i) without premagnetisation, i.e., the bulk is in the virgin state when the crossed field is applied and (ii) with premagnetisation. The distribution of the superconducting current within the bulk in each of these cases is shown in figure 4.20, for a crossed-field of 0.1 T. An asymmetry arises when the

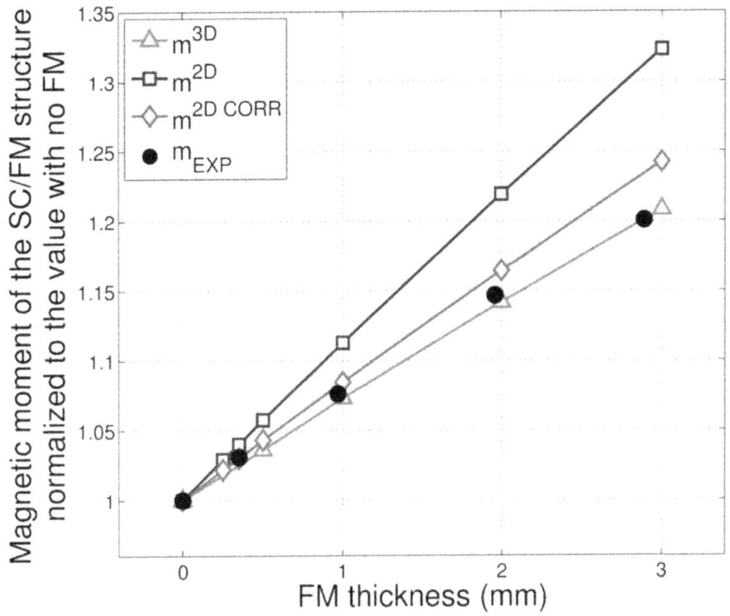

Figure 4.19. Magnetic moment normalised to the value with no ferromagnet: measurements (m_{EXP}), modelled in 3D (m^{3D} = $m_{TOT}{}^{3D}/m_{TOT}{}^{3D}$ (0 mm)) and in 2D (m^{2D} = $m_{TOT}{}^{2D}/m_{TOT}{}^{2D}$ (0 mm)) for the whole hybrid structure. In 2D and 3D modelling, it is obtained by addition of the separated m_{SC} and m_{FM} contributions. The superscript CORR refers to the corrected values obtained by the addition of the m_{SC} contribution and ¾ of the m_{FM} contribution modelled in 2D. © IOP Publishing. Reproduced by permission of IOP Publishing from [73]. All rights reserved.

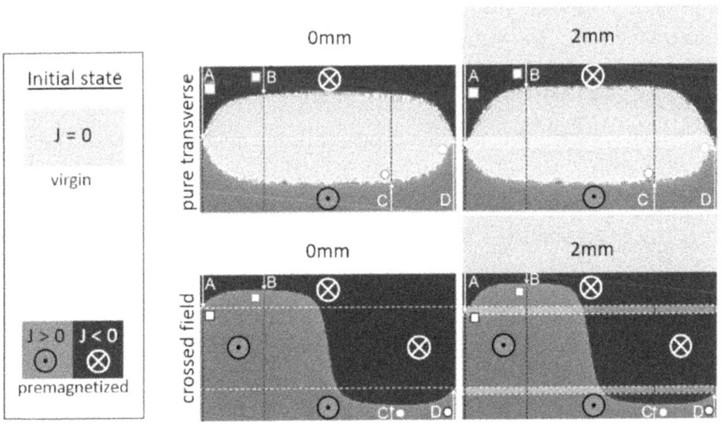

Figure 4.20. Distribution of the superconducting currents in the bulk in the case of the 'virgin' state (top) and the 'premagnetised' state (bottom). The left side illustrates the case of the 0 mm structure (bulk only, no ferromagnet) and the right side illustrates the case of the 2 mm structure where the 2 mm ferromagnetic layer is located above the bulk. All distributions correspond to a transverse magnetic field increased up to $\mu_0 H_{app}$ = 0.1 T. © IOP Publishing. Reproduced by permission of IOP Publishing from [73]. All rights reserved.

ferromagnet is present and the authors conclude that, because the ferromagnet is subjected to a concentration of magnetic flux, a higher magnetic field exists at the corner of the bulk near the ferromagnet, which enhances the penetration of the current in this region. Since the net current must be equal to zero, the penetration inside the bulk on the other side of the bulk is less. The authors find that (i) the ferromagnetic layer has a beneficial effect on the magnetic moment due to a crossed-field, (ii) there exists a threshold thickness for the ferromagnetic layer, above which this improvement becomes marginal and (iii) this threshold thickness is related to the saturation magnetisation of the ferromagnet. Placing the ferromagnetic layer around the sides of the bulk ('coax-x mm' configuration) also reduces the crossed-field effect, but has an additional advantage of slightly increasing the local magnetic field above and below the structure, while shielding its sides.

4.2.3 Hollow bulk cylinders and tubes for shielding

In addition to acting as TFMs that can provide high magnetic fields, hollow, cylindrical or tube-shaped bulk superconductors are also able to shield magnetic fields in a more efficient way than an equivalent ferromagnetic cylinder [76–78]. This shielding property has also been exploited in bulk superconductor reluctance machines [79, 80], where the magnetic field along the direct axis is enhanced by reinforcing the low reluctance path provided by the ferromagnetic material in the rotor (see section 1.5.2.2).

In [77], the authors present an analysis of the magnetic shielding properties of a polycrystalline Bi-2223 superconducting tube subjected to AC and DC axial magnetic fields. A numerical model based on [25, 26] is again used to simulate the experimental results. The tube can effectively shield an axial field up to a threshold axial field, B_{lim}, that is determined by the length of the tube and its mean radius, but also the exact $J_c(B)$ dependence of the superconductor. For a decreasing tube length, B_{lim} can be severely reduced due to demagnetisation effects. Two mechanisms for the field penetration in the superconductor are described: one from the external surface and one from the opening ends of the tube; the latter being suppressed for long tubes. Finally, the shielding factor is increased for increasing frequency, which is explained in terms of the power law, shown in equation (2.10) in section 2.2.1.3.2. However, in this study, the effect of heat generation due to the motion of vortices was not examined, which may have a significant influence as described in [10].

In [78], this study is extended to analyse the effect of a cap on the ends of the tube, effectively closing the tube, and for tubes of shorter length, this has a noticeable, positive effect on the maximum shieldable field and the uniformity of the shielding. In addition, the effect of a non-superconducting joint and of inhomogeneous superconducting properties is investigated, which both can act to reduce the magnitude and location of maximum field attenuation.

In [81], the authors consider magnetic shielding of various geometries of bulk semi-closed superconducting cylinders subjected to axial and transverse fields. A 2D numerical model based on the A-ϕ formulation is used to simulate axial field shielding, but a 3D model is required for the transverse field configuration. Both models are implemented in the open-source solver GetDP mentioned earlier in

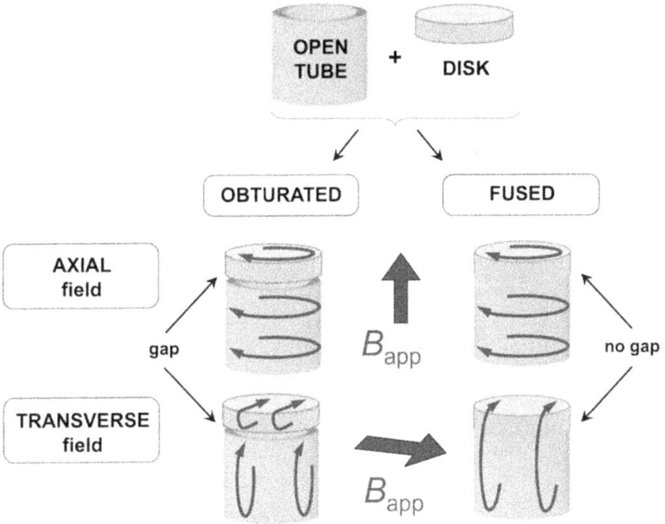

Figure 4.21. Schematic diagram of two semi-closed geometries of magnetic shields combining an open tube and a superconducting disc. In the 'obturated' configuration the disc plate is placed above one opening of the tube. In the 'fused' configuration, the disc and the open tube are grown in the same process, resulting in a 'cup-shaped' sample with no gap between the tube and the closing. The supercurrent paths in the superconducting parts are drawn schematically when the magnetic field is applied either parallel or perpendicular to the axis of the shields.

section 4.2.2. Figure 4.21 shows a schematic diagram of two semi-closed geometries of magnetic shields combining an open tube and a superconducting disc. In the 'obturated' configuration, the disc plate is placed above one opening of the tube. In the 'fused' configuration, the disc and the open tube are grown in the same process, resulting in a 'cup-shaped' sample with no gap between the tube and the closing. The supercurrent paths in the superconducting parts are drawn schematically when the magnetic field is applied either parallel or perpendicular to the axis of the shields. The magnetic shield geometries investigated in [81] are shown in figure 4.22. The reference geometry is a hollow superconducting tube of internal radius, R_{in}, height h and wall thickness e. Geometric dimensions, $R_{in} = 10$ mm, $h = 30$ mm and $e = 3$ mm, are assumed, which are typical of bulk melt-textured YBCO cylinders fabricated by buffer-aided top-seeded melt growth [82, 83] and have been studied in previous experimental works [84, 85].

Figure 4.23 shows the distribution of the resulting axial shielding factor, SF $= B_{app}/|B_z|$, for an applied field amplitude $B_{app} = 3$ mT, which is much less than the full penetration field, B_{pen}, calculated analytically in [81] as 29.8 mT. In this regime, for the superconducting tube, the penetration occurs mainly from the open ends and not through the superconducting walls. The grey zones are characterised by an axial shielding factor SF > 1000. The shielding factor for the double tube is found to be higher than the single tube. The other configurations all show similar behaviour to the single/double tube at the bottom (where there is an opening), but there is almost no penetration from the top, even with a small gap.

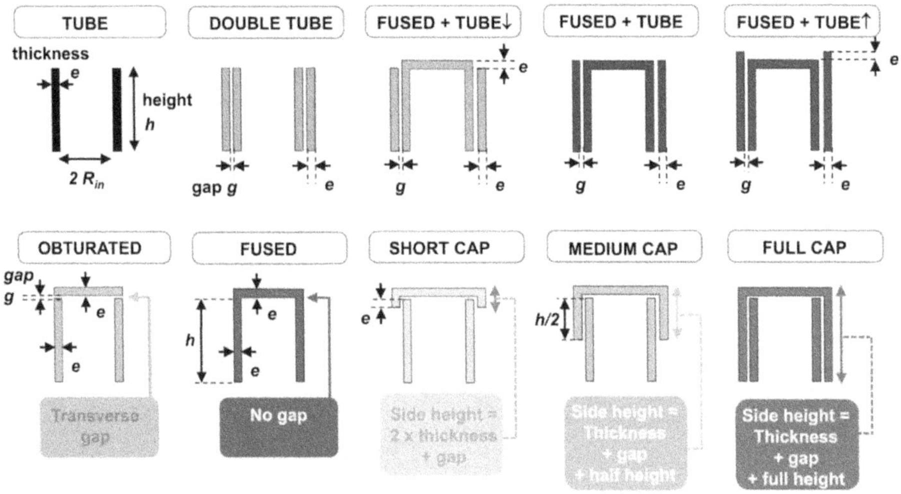

Figure 4.22. Schematic illustration of the axisymmetric geometries investigated numerically in [81]. The assumed geometric dimensions are $R_{\text{in}} = 10$ mm, $h = 30$ mm and $e = 3$ mm. All gaps between superconducting parts are equal to $g = 1$ mm. All superconducting parts are characterised by the same superconducting properties. © IOP Publishing. Reproduced by permission of IOP Publishing from [81]. All rights reserved.

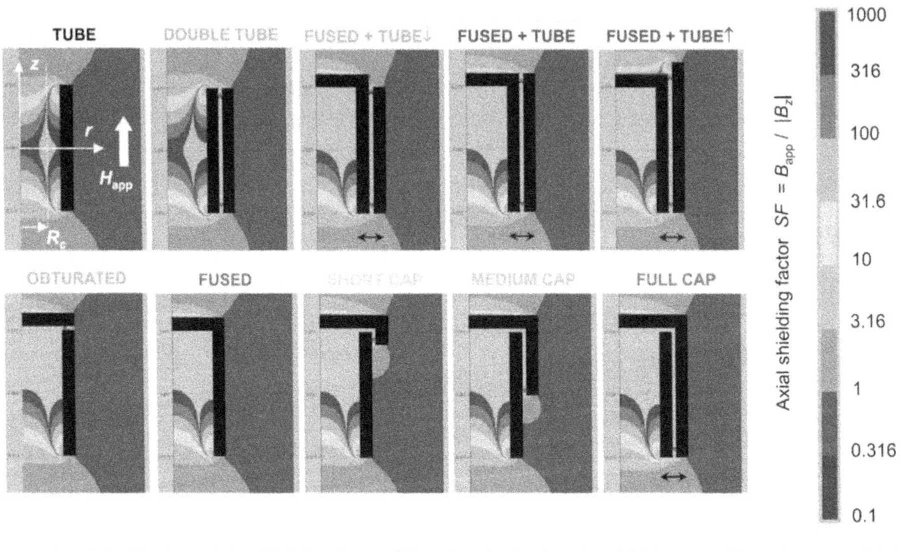

Figure 4.23. Distribution of the shielding factor $\text{SF} = B_{\text{app}}/|B_z|$ when the shields are subjected to an axial field $B_{\text{app}} \parallel z$. The amplitude of the applied field is $B_{\text{app}} = 3$ mT. The zones in grey are characterised by an axial shielding factor SF > 1000. © IOP Publishing. Reproduced by permission of IOP Publishing from [81]. All rights reserved.

Figure 4.24 shows the magnetic field dependence of the axial shielding factor at the centre of the tube ($r = z = 0$ mm) when the applied field is increased from 3 to 30 mT. The highest SF is observed for the 'fused + tube' and 'full cap' configurations, while the 'tube' and 'double tube' configurations have the lowest SF. There

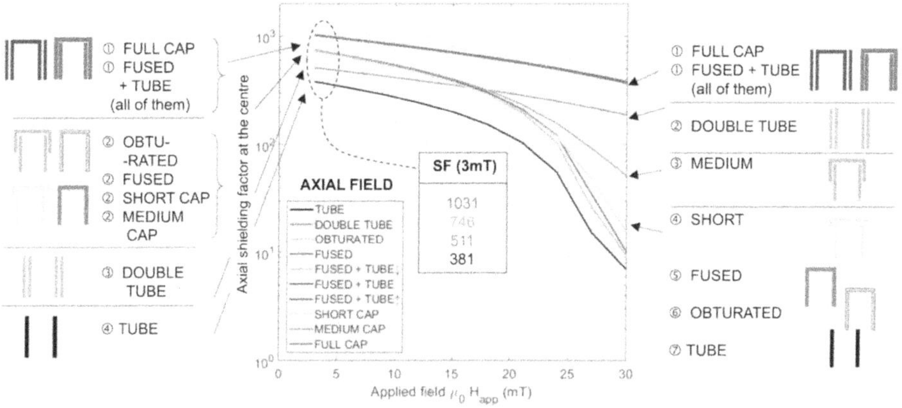

Figure 4.24. Magnetic field dependence of the shielding factor SF = $B_{app}/|B_z|$ at the centre of the inner tube ($r = z = 0$) when the magnetic shields are subjected to an axial field $B_{app} \parallel z$. The inset shows the values of the shielding factors measured at B_{app} = 3 mT for the four groups of shields.

is also a cross-over for the 'double tube' and the other configurations between low and high fields. For low fields, a configuration closed at one end is preferred, and for higher fields, penetration through the lateral walls becomes predominant, so thick or double walls are preferred.

Figure 4.25 shows the magnetic field dependence of the transverse shielding factor, SF = $B_{app}/|B_y|$, either at the centre of the tube ($r = z = 0$ mm) or at the top side ($r = 0$ mm, $z = h/2$). The shielding factors at the centre of the tube (figure 4.25(a)) are fairly low (SF < 20) for all configurations, but the highest SF values are obtained for the shields combining a tube and cap coaxially, i.e., the 'full cap' and 'fused + tube' configurations. Significantly higher SF values are found near the closed extremities of the semi-closed shields (see figure 4.25(b)), with the best shielding properties observed for the 'fused + tube' configuration; in particular, the geometry in which the height of the surrounding tube extends slightly over that of the cup, i.e., the 'fused + tube↑' configuration. The numerical modelling provides a useful visualisation of the magnetic field lines and allows optimisation of shield design to maximise the shielding factor.

Finally, in [86], the authors carry out a numerical investigation of the shielding properties of hybrid ferromagnetic-superconductor shielding tubes, which had previously been studied mostly experimentally [87–89]. The addition of a ferromagnetic cylinder either inside or outside the superconducting cylinder, as shown in figure 4.26, can substantially improve the shielding limit, which is maximised using the 'Ferro Out' configuration, i.e., the ferromagnetic tube is placed outside the HTS tube. With the assistance of numerical simulations, which include the electrical properties of the superconductor and saturation of the ferromagnet, the shielding effect for the 'Ferro Out' configuration is decomposed into two processes: attenuation from the ferromagnetic layer (independent of the HTS layer) and a decrease in magnetic flux density in the wall of the HTS tube, according to its own shielding

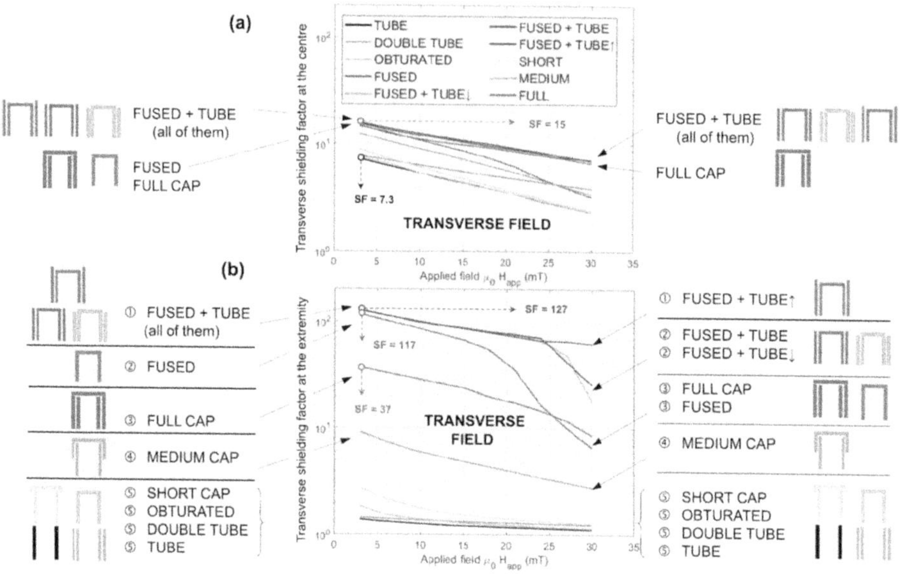

Figure 4.25. Magnetic field dependence of the shielding factor SF = $B_{app}/|B_y|$ when the magnetic shields are subjected to a transverse field $B_{app} \parallel y$ ($\perp z$). Particular values of the shielding factors measured at $B_{app} = 3$ mT are also shown. (a) Shielding factor at the centre of the inner tube ($r = z = 0$ mm). (b) Shielding factor at the extremity of the inner tube ($r = 0$ mm, $z = h/2 = 15$ mm).

factor. The mechanism for the 'Ferro In' configuration is slightly different in that the presence of the ferromagnetic tube improves the magnetic field decay in the HTS tube. The height of the ferromagnetic tube also plays an important role, which can be optimised using numerical methods.

4.2.4 Hybrid trapped field magnet lens

In [90], a new concept of a hybrid trapped field magnet lens (HTFML) is proposed. The HTFML exploits the 'vortex pinning effect' of an outer bulk superconducting cylinder, which is magnetised as a TFM using FC magnetisation, and the 'diamagnetic shielding effect' of an inner bulk magnetic lens to generate a concentrated magnetic field in the bore of the lens higher than the trapped field from the TFM. This requires that, during the magnetisation process, the outer cylinder is in the normal state ($T > T_c$) and the inner lens is in the superconducting state ($T < T_c$) when the external magnetising field is applied, followed by cooling to an appropriate operating temperature such that both parts are in the superconducting state, then removing the external field. Figure 4.27(a) shows the time sequence of conventional FC magnetisation of a bulk cylinder as a TFM and figure 4.27(b) shows the time sequence of the magnetisation process for a conventional bulk superconducting magnetic lens, for which ZFC magnetisation is used. Such lenses have been investigated experimentally: a concentrated field of $B_c = 12.42$ T has been achieved at 20 K for a background field of $B_{app} = 8$ T using a bulk GdBCO magnetic

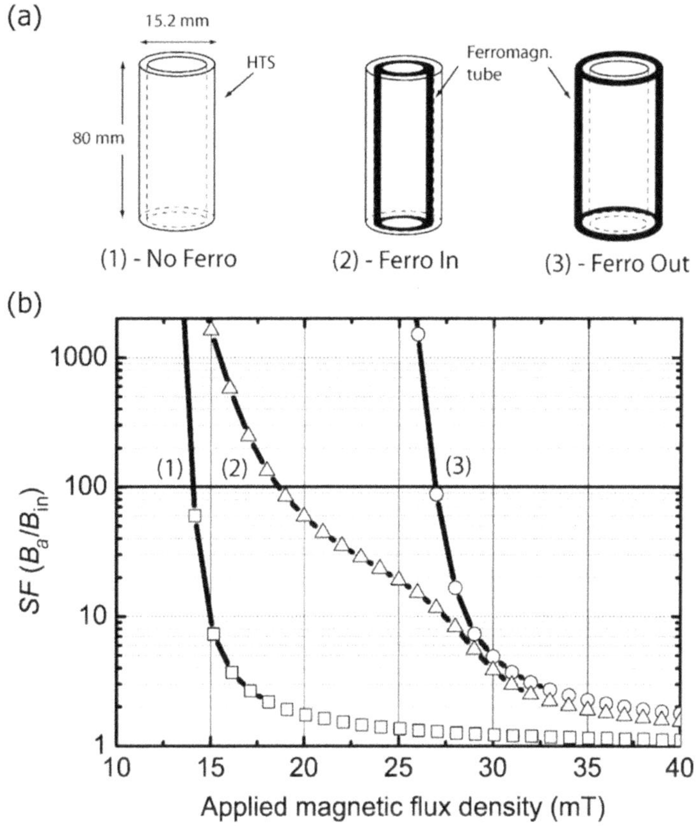

Figure 4.26. (a) Configurations of the ferromagnet–superconductor tubes considered in [86]. The ferromagnetic tube (black) is placed either (2) inside ('Ferro In') or (3) outside ('Ferro Out') the HTS tube. (b) The shielding factor, SF, as a function of applied magnetic flux density for the three configurations. © 2010 IEEE. Reprinted, with permission, from [86].

lens [91] and B_c = 30.4 T has been achieved at the centre of the lens in a higher background field B_{app} = 28.3 T [92]. Using a bulk MgB_2 magnetic lens, a concentrated field of B_c = 2.18 T at 4.2 K has also been achieved for a background field B_{app} = 1 T [93]. However, in these conventional lenses, the external magnet providing the background field must be operated continuously or the magnetic lens effect disappears. A significant advantage of the HTFML is that it can reliably generate a magnetic field at the centre of the lens higher than the trapped field of the bulk cylinder TFM and the external magnetising field, even after this field is removed.

The HTFML is explored for two potential cases in [90]: (1) exploiting the difference in T_c of two different bulk materials ('case-1'), e.g., MgB_2 (T_c = 39 K) and GdBCO (T_c = 92 K), and (2) using the same material for the whole HTFML, e.g., GdBCO, but utilising individually controlled cryostats, the same cryostat with different cooling loops or coolants, or heaters that keep the outer bulk cylinder at a temperature above T_c to achieve the same desired effect. The numerical modelling

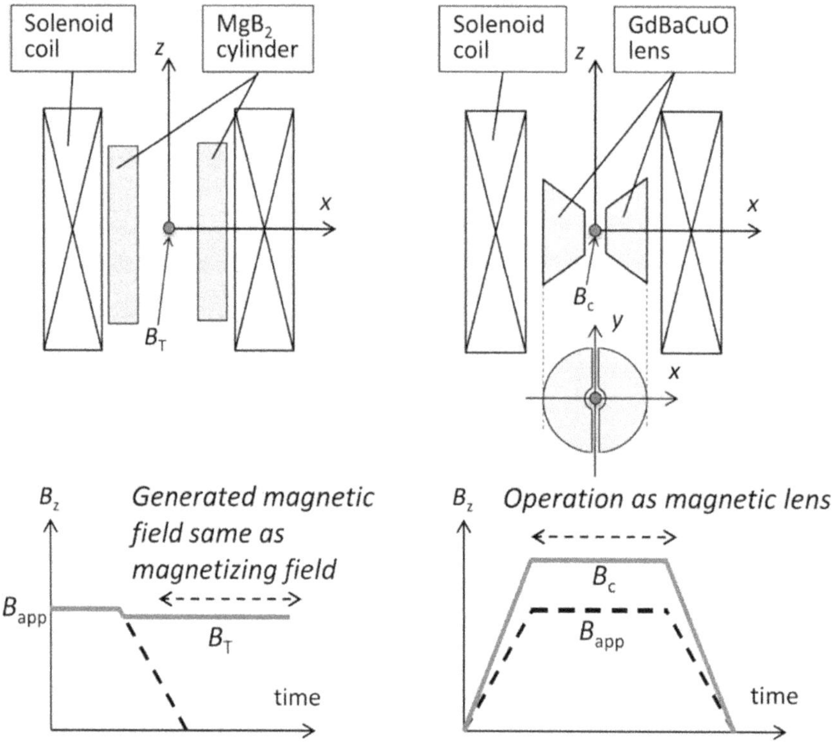

Figure 4.27. Time sequence of (a) conventional magnetising process of FC magnetisation of a bulk superconducting cylinder to utilise it as a TFM and (b) ZFC magnetisation of a conventional superconducting bulk magnetic lens.

framework is based on the *A*-*V* formulation described in [94]. The *E*–*J* power law is used with $n = 20$ for the GdBCO bulk and $n = 100$ for the MgB_2 bulk. The assumed $J_c(B)$ characteristics of the GdBCO and MgB_2 bulks are based on practical fitted values using equations (2.8) and (2.9), respectively. The numerical model for 'case-1' is shown in figure 4.28, including the assumed dimensions of the MgB_2 cylinder, GdBCO magnetic lens and solenoid magnetising coil, and the time step sequence of the magnetising process is shown in figure 4.29 for (a) the temperature, T, and (b) the external magnetising field, B_{ex}. For 'case-1,' the magnetising process is as follows:

(1) The bulk MgB_2 cylinder and the bulk GdBCO lens are cooled from 100 K to $T_H = 40$ K, which is higher than the superconducting transition temperature of MgB_2 (39 K), but lower than that of GdBCO (92 K). In this stage, the MgB_2 cylinder is in the normal state and the GdBCO lens is in the superconducting state (step 0).
(2) The external magnetic field, B_{ex}, is ramped up linearly at 0.222 T min^{-1} over five steps (steps 1–5) up to B_{app}, which is equivalent to ZFC magnetisation of the GdBCO lens. The magnetic field, essentially higher than B_{app} because of the shielding effect by the magnetic lens, completely penetrates the MgB_2 cylinder and the magnetic field is concentrated at the centre of the lens.

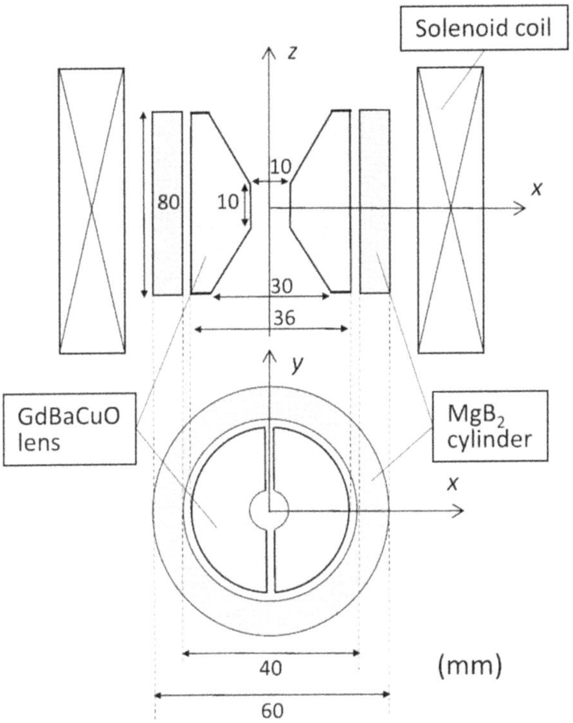

Figure 4.28. Numerical model for 'case-1,' including the assumed dimensions of the MgB_2 cylinder, GdBCO magnetic lens and solenoid magnetising coil. The MgB_2 cylinder is replaced by a GdBCO cylinder in 'case-2.'

(3) The temperatures of both the MgB_2 cylinder and GdBCO lens are then decreased to $T_L = 20$ K, which is lower than the T_c of MgB_2.
(4) B_{ex} is decreased linearly at 0.222 T min^{-1} over five steps (steps 6–10) down to zero. During this process, the MgB_2 cylinder is magnetised by FC magnetisation and magnetic flux is trapped in the cylinder. The magnetic field concentration effect slightly decreases due to the decrease of external field. However, a magnetic field at the centre of the magnetic lens still remains due to the existence of the trapped field in the MgB_2 cylinder.
(5) As a result, HTFML can reliably generate a magnetic field higher than B_t of the single cylindrical TFM and B_{app}, even after $B_{ex} = 0$ T.

In 'case-2,' in which both the superconducting cylinder and magnetic lens are made from bulk GdBCO, (1) and (2) above are changed as follows, labelled as (1') and (2'):

(1') The bulk GdBCO cylinder is maintained at 100 K (dashed orange line in figure 4.29) and the bulk GdBCO magnetic lens is cooled to $T_H = 40$ K (blue line in figure 4.29). Hence, the bulk GdBCO cylinder is in the normal state and the bulk GdBCO lens is in the superconducting state (step 0).

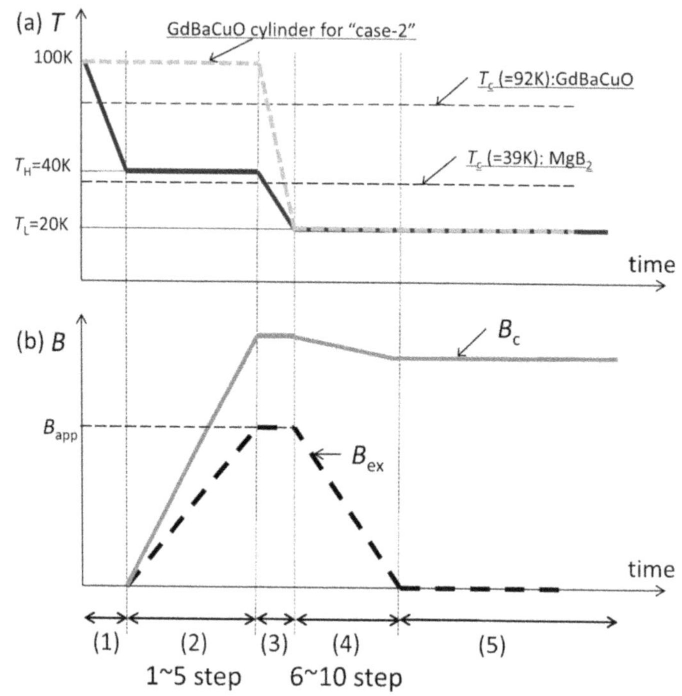

Figure 4.29. Time step sequence of (a) the temperature, T, and (b) the external field, B_{ex}, and concentrated magnetic field, B_c, at the centre of the magnetic lens for 'case-1.' In 'case-2,' the temperature of the GdBCO lens follows the blue line and the GdBaCuO cylinder follows the dashed orange line in the upper panel.

(2') The external magnetic field, B_{ex}, is increased linearly over five steps (steps 1–5) up to B_{app}, and the magnetic field, essentially equal to B_{app}, completely penctrates the GdBCO cylinder, but the GdBCO lens is magnetised by ZFC magnetisation.

Figure 4.30 shows the time step dependence of the magnetic field profile along the x-direction across the centre of the lens during (a) the ascending stage and (b) the descending stage of the magnetisation process for 'case-1' for $B_{app} = 3$ T. During the ascending stage (steps 0–5, see figure 4.30(a)), the concentrated magnetic field, B_c, at the centre of the lens is enhanced with increasing B_{ex}, due to the diamagnetic shielding effect and there is little penetration of the field into the lens itself. During the descending stage (figure 4.30(b)), B_c decreases somewhat with decreasing B_{ex}, but becomes stable after the external field decreases to zero and a magnetic field is trapped in the MgB_2 cylinder. At the final step (step 10), $B_c = 4.73$ T, which is significantly higher than the applied field, 3 T. Figure 4.31 shows B_c as a function of B_{ex} for $B_{app} = 3$ T, extracted from figure 4.30, including the situation where the lens is not present, i.e., the MgB_2 cylinder only, highlighting the difference between the TFM only and the HTFML.

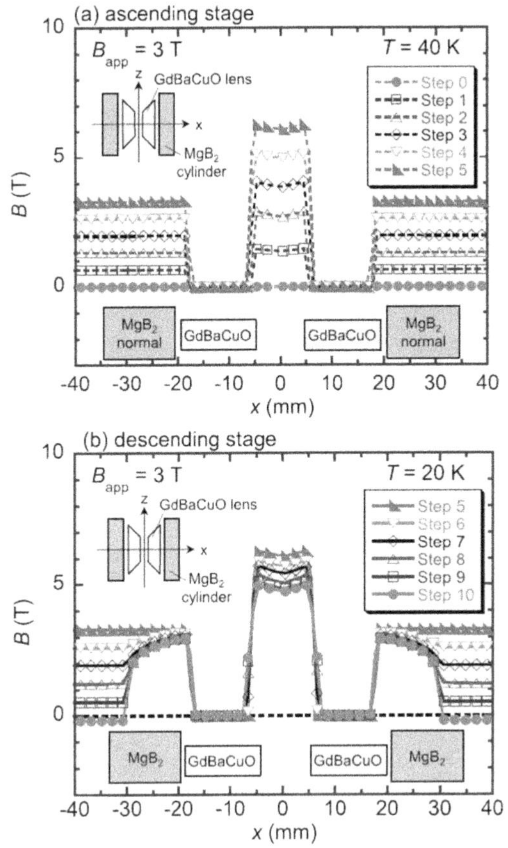

Figure 4.30. Time step dependence of the magnetic field profile along the x-direction across the centre of the lens during (a) the ascending stage and (b) the descending stage of the magnetisation process for 'case-1' for $B_{app} = 3$ T.

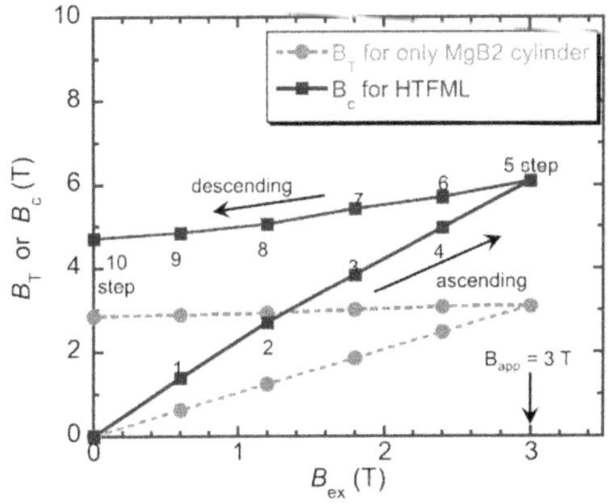

Figure 4.31. Concentrated magnetic field, B_c, at the centre of the magnetic lens as a function of the external field, B_{ex}, for $B_{app} = 3$ T. The trapped field, B_t, at the centre of the MgB_2 cylinder for the case without the lens is also shown.

Figure 4.32 shows the time step dependence of the magnetic field profile along the *x*-direction across the centre of the lens during (a) the ascending stage and (b) the descending stage of the magnetisation process for 'case-2' for B_{app} = 10 T. In this case, B_c is as high as 16 T at the centre of the lens at step 5 and settles to a final value of 13.5 T at the end of the magnetisation process (step 10). Figure 4.33 shows B_c as a function of B_{ex} for B_{app} = 10 T, extracted from figure 4.32, as well as the results for B_{app} = 3 and 6 T. Magnetic field concentration ratios, B_c/B_{app}, of 1.7, 1.5 and 1.33 are achieved for B_{app} of 3, 6 and 10 T, respectively.

Such a device could become a standard method for trapped field enhancement in several practical bulk superconductor applications; for example, for enhancing the magnetic field in the bore of a bulk superconducting NMR/MRI system to improve its resolution.

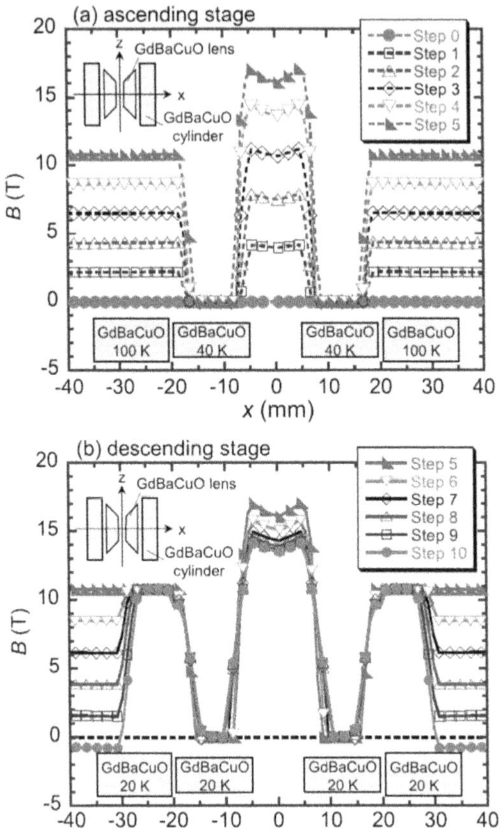

Figure 4.32. Time step dependence of the magnetic field profile along the *x*-direction across the centre of the lens during (a) the ascending stage and (b) the descending stage of the magnetisation process for 'case-2' for B_{app} = 10 T.

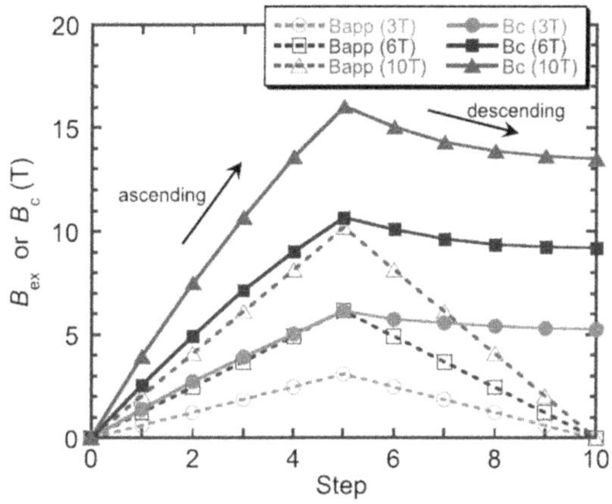

Figure 4.33. Concentrated magnetic field, B_c, at the centre of the magnetic lens as a function of the external field, B_{ex}, for B_{app} = 3, 6 and 10 T for 'case-2.' The external magnetising field, B_{ex}, generated by the solenoid magnetising coil is also shown for each B_{app}. © IOP Publishing. Reproduced by permission of IOP Publishing from [90]. All rights reserved.

References

[1] Tinkham M 1996 *Introduction to Superconductivity* (New York: Dover), p 179

[2] Anderson P W 1962 *Phys. Rev. Lett.* **9** 309–11

[3] Caplin A D, Cohen L F, Perkins G K and Zhukov A A 1994 *Supercond. Sci. Technol.* **7** 412–22

[4] Yamasaki H and Mawatari Y 2000 *Supercond. Sci. Technol.* **13** 202–8

[5] Cardwell D A *et al* 2005 *Supercond. Sci. Technol.* **18** S173–9

[6] Vanderbemden P *et al* 2007 *Phys. Rev.* B **75** 174515

[7] Baghdadi M, Ruiz H S and Coombs T A 2014 *Appl. Phys. Lett.* **104** 232602

[8] Vanderbemden P *et al* 2007 *Supercond. Sci. Technol.* **20** S174–83

[9] Campbell A M 2014 *Supercond. Sci. Technol.* **27** 124006

[10] Zou J, Ainslie M D, Hu D and Cardwell D C 2015 *IEEE Trans. Appl. Supercond.* **25** 4900505

[11] Qiu M *et al* 2005 *IEEE Trans. Appl. Supercond.* **15** 3172–5

[12] Qiu M *et al* 2004 *IEEE Trans. Appl. Supercond.* **14** 1989–91

[13] Ma G T *et al* 2008 *Supercond. Sci. Technol.* **21** 065020

[14] Ogawa J *et al* 2003 *Physica* C **386** 26–30

[15] Yamagishi K, Ogawa J, Tsukamoto O, Murakami M and Tomita M 2003 *Physica* C **392–396** 659–63

[16] Tsukamoto O *et al* 2005 *J. Mater. Process. Technol.* **161** 52–7

[17] Srpcic J *et al* 2019 *Supercond. Sci. Technol.* **32** 035010

[18] Kapolka M *et al* 2018 *IEEE Trans. Appl. Supercond.* **28** 6801405

[19] Pardo E and Kapolka M 2017 *J. Comput. Phys.* **344** 339–63

[20] Zushi Y *et al* 2004 *Physica* C **412** 708–13

[21] Berger K *et al* 2007 *IEEE Trans. Appl. Supercond.* **17** 3028–31

[22] Zushi Y, Asaba I, Ogawa J, Yamagishi K and Tsukamoto O 2005 *Cryogenics* **45** 17–22

[23] Vanderbemden P *et al* 2010 *Supercond. Sci. Technol.* **23** 075006

[24] Laurent P *et al* 2010 *Supercond. Sci. Technol.* **23** 124004
[25] Brandt E H 1998 *Phys. Rev.* B **58** 6506–22
[26] Brandt E H 1996 *Phys. Rev.* B **54** 4246–64
[27] Prigozhin L 1996 *J. Comput. Phys.* **129** 190–200
[28] Fagnard J F *et al* 2009 *Supercond. Sci. Technol.* **22** 105002
[29] Yamagishi K, Tsukamoto O, Ogawa J and Miyagi D 2011 *IEEE Trans. Appl. Supercond.* **21** 3320–4
[30] Yamagishi K, Ogawa J and Tsukamoto O 2014 *J. Phys. Conf. Ser.* **507** 032060
[31] Yamagishi K, Ogawa J and Tsukamoto O 2015 *IEEE Trans. Appl. Supercond.* **25** 5201905
[32] Fukui S *et al* 2006 *IEEE Trans. Appl. Supercond.* **16** 143–6
[33] Patel A and Glowacki B A 2012 *Supercond. Sci. Technol.* **25** 125015
[34] Patel A *et al* 2013 *Appl. Phys. Lett.* **102** 102601
[35] Patel A, Hopkins S C and Glowacki B A 2013 *Supercond. Sci. Technol.* **26** 032001
[36] Selva K and Majkic G 2013 *Supercond. Sci. Technol.* **26** 115006
[37] Selva K, Li X-F and Majkic G 2015 *IEEE Trans. Appl. Supercond.* **25** 6605005
[38] Cavallin T, Young E A, Beduz C, Yang Y and Giunchi G 2007 *IEEE Trans. Appl. Supercond.* **17** 2770–3
[39] Fujishiro H, Ujiie T, Mochizuki H, Yoshida T and Naito T 2015 *IEEE Trans. Appl. Supercond.* **25** 6800104
[40] Noudem J G, Meslin S, Harnois C, Chateigner D and Chaud X 2004 *Supercond. Sci. Technol.* **17** 931–6
[41] Chaud X, Isfort D, Porcar L and Tournier R 2005 *J. Eur. Ceram. Soc.* **25** 2955–8
[42] Meslin S *et al* 2006 *Supercond. Sci. Technol.* **19** S585–9
[43] Lousberg G P *et al* 2011 *Supercond. Sci. Technol.* **24** 035008
[44] Tomita M and Murakami M 2003 *Nature* **421** 517–20
[45] Yokoyama K, Oka T, Kondo N and Hosaka S 2013 *IEEE Trans. Appl. Supercond.* **23** 8201204
[46] Baskys A *et al* 2014 *J. Phys. Conf. Ser.* **507** 012003
[47] Lousberg G *et al* 2009 *Supercond. Sci. Technol.* **22** 055005
[48] Majoros M, Glowacki B A and Campbell A M 2001 *IEEE Trans. Appl. Supercond.* **11** 2780–3
[49] Majoros M, Sumption M D and Collings E W 2009 *IEEE Trans. Appl. Supercond.* **19** 3352–5
[50] Majoros M, Glowacki B A and Campbell A M 2000 *Physica* C **338** 251–62
[51] Kovac P *et al* 2003 *Supercond. Sci. Technol.* **16** 1195–201
[52] Gomory F, Vojenciak M, Pardo E and Souc J 2009 *Supercond. Sci. Technol.* **22** 034017
[53] Kruger P, Grilli F, Vojenciak M, Zermeno V M R, Demencik E and Farinon S 2013 *Appl. Phys. Lett.* **102** 202601
[54] Gomory F, Vojenciak M, Pardo E, Solovyov M and Souc J 2010 *Supercond. Sci. Technol.* **23** 034012
[55] Safran S, Gomory F and Gencer A 2010 *Supercond. Sci. Technol.* **23** 085011
[56] Liu G, Zhang G, Jing L and Yu H 2017 *Supercond. Sci. Technol.* **30** 125014
[57] Vojenciak M, Souc J and Gomory F 2011 *Supercond. Sci. Technol.* **24** 075001
[58] Wen H *et al* 2009 *IEEE Trans. Appl. Supercond.* **19** 1652–5
[59] Xing Y Q *et al* 2016 *IEEE Trans. Appl. Supercond.* **26** 5501305
[60] Pardo E, Souc J and Vojenciak M 2009 *Supercond. Sci. Technol.* **22** 075007
[61] Ainslie M D, Yuan W and Flack T J 2013 *IEEE Trans. Appl. Supercond.* **23** 4700104

[62] Ainslie M D, Hu D, Zou J and Cardwell D A 2015 *IEEE Trans. Appl. Supercond.* **25** 4602305
[63] Hu D *et al* 2017 *IEEE Trans. Appl. Supercond.* **27** 4701506
[64] Campbell A M and Cardwell D A 1997 *Cryogenics* **37** 567–75
[65] Granados X, Torner M, Puig T and Obradors X 2007 *IEEE Trans. Appl. Supercond.* **17** 1629–32
[66] Philippe M P *et al* 2014 *Physica* C **502** 20–30
[67] Campbell A M 2009 *Supercond. Sci. Technol.* **22** 034005
[68] Campbell A M 2007 *Supercond. Sci. Technol.* **20** 292–5
[69] Ainslie M D *et al* 2011 *Supercond. Sci. Technol.* **24** 045005
[70] Ainslie M D, Flack T J and Campbell A M 2012 *Physica* C **472** 50–6
[71] Nguyen D N, Ashworth S P, Willis J O, Sirois F and Grilli F 2010 *Supercond. Sci. Technol.* **23** 025001
[72] Philippe M P *et al* 2015 *Supercond. Sci. Technol.* **28** 095008
[73] Fagnard J F *et al* 2016 *Supercond. Sci. Technol.* **29** 125004
[74] Dular P and Geuzaine C 2006 GetDP reference manual: the documentation for GetDP, a general environment for the treatment of discrete problems (http://geuz.org/getdp)
[75] Dular P 2001 *Scientific Computing in Electrical Engineering* vol 18 (Berlin: Springer) pp 367
[76] Gyoda M, Torii S and Ebihara D 1998 *Adv. Supercond.* XII (Tokyo: Springer) pp 1353–6
[77] Denis S *et al* 2007 *Supercond. Sci. Technol.* **20** 192–201
[78] Denis S, Dirickx M, Vanderbemden P, Ausloos M and Vanderheyden B 2007 *Supercond. Sci. Technol.* **20** 418–27
[79] Barnes G J, McCulloch M D and Dew-Hughes D 2000 *Supercond. Sci. Technol.* **13** 875–78
[80] Kovalev L *et al* 2002 *Supercond. Sci. Technol.* **15** 817–22
[81] Fagnard J F, Vanderheyden B, Pardo E and Vanderbemden P 2019 *Supercond. Sci. Technol.* **32** 074007
[82] Kumar N D, Shi Y, Zhai W, Dennis A R, Durrell J H and Cardwell D A 2015 *Cryst. Growth Des.* **15** 1472–80
[83] Namburi D K *et al* 2016 *Supercond. Sci. Technol.* **29** 034007
[84] Wera L *et al* 2017 *IEEE Trans. Appl. Supercond.* **27** 6800305
[85] Wera L *et al* 2019 *IEEE Trans. Appl. Supercond.* **29** 6801109
[86] Lousberg G P *et al* 2010 *IEEE Trans. Appl. Supercond.* **20** 33–41
[87] Itoh M, Ohyama T, Hoshino K, Ishigaki H and Minemoto T 1993 *IEEE Trans. Appl. Supercond.* **3** 181–4
[88] Itoh M, Mori K and Minemoto T 1996 *IEEE Trans. Magn.* **32** 2605–8
[89] Mori K, Minemoto T and Itoh M 1997 *IEEE Trans. Appl. Supercond.* **7** 378–81
[90] Takahashi K, Fujishiro H and Ainslie M D 2018 *Supercond. Sci. Technol.* **31** 044005
[91] Zhang Z Y, Matsumoto S, Teranishi R and Kiyoshi T 2012 *Supercond. Sci. Technol.* **25** 115012
[92] Choi S *et al* 2012 *J. Appl. Phys.* **111** 07E728
[93] Zhang Z Y *et al* 2012 *Supercond. Sci. Technol.* **25** 025009
[94] Fujishiro H and Naito T 2011 *Supercond. Sci. Technol.* **23** 105021

Mark Ainslie and Hiroyuki Fujishiro

Appendix A

Thermal properties of bulk superconductors

A.1 Introduction

The thermal conductivity of a material determines its ability to conduct heat. Such data for bulk superconducting materials are necessary to calculate heat transfer and diffusion for the bulk during the magnetisation process, especially for pulsed field magnetisation. The heat Q_{int} transferred from a higher temperature (T_{H}) side to a lower temperature side (T_{L}) is in general given by

$$Q_{\text{int}} = \frac{S}{L}\int_{T_{\text{L}}}^{T_{\text{H}}} \kappa(T)\mathrm{d}T \tag{A.1}$$

where $\kappa(T)$ is the thermal conductivity, L the length and S the cross-section of the bulk material. The thermal conductivity is given by

$$\kappa = \rho C\alpha \tag{A.2}$$

where ρ is the mass density, C is the specific heat and α is the thermal diffusivity.

A.2 Experimental procedure

A.2.1 Thermal conductivity

Thermal conductivity, $\kappa(T)$, is usually measured by a steady-state heat flow method. Here, we introduce the basic measuring technique of $\kappa(T)$ for bulk superconductors in our laboratory. The experimental setup for the $\kappa(T)$ measurement is shown schematically in figure A1. One end of the sample is soldered to the copper block (sample stage) on the cold stage of a Gifford-McMahon (GM) cycle helium refrigerator. A small metal chip resistor of 1 kΩ is used as a heater, which is

doi:10.1088/978-0-7503-1332-2ch5

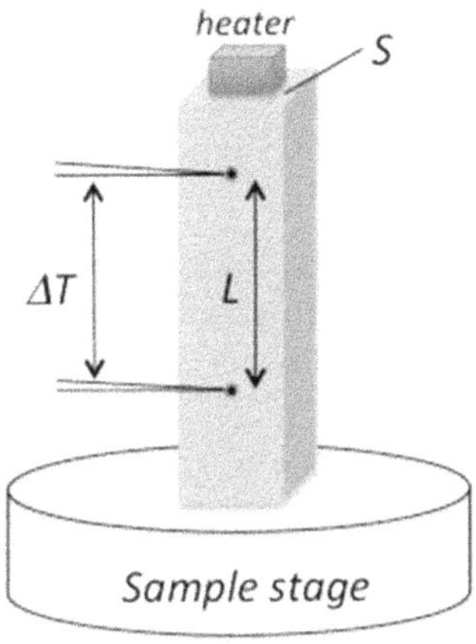

Figure A1. Experimental setup for the measurement of thermal conductivity, $\kappa(T)$.

adhered to the other end of the sample by GE7031 varnish. κ is obtained using the following relation,

$$\kappa = \frac{Q}{\Delta T}\frac{L}{S} \tag{A.3}$$

where Q is the applied heat from the heater, ΔT is the temperature difference, L is the distance between two chromel-constantan (or Au(Fe)-chromel) thermocouples (76 μm in diameter), and S is the cross-section of the sample. ΔT is maintained within 0.5–0.8 K by controlling the heater power automatically using a computer. To reduce the radiation loss, $\kappa(T)$ was measured in a vacuum with radiation shielding.

A.2.2 Thermal dilatation

The thermal dilatation, dL/L, is also an important physical property, because the change in the length results in a mechanical stress to the supporting material. Thermal dilatation, normalised by the sample length L at 300 K, $\Delta L(T)/L(300\text{ K}) = (L(T) - L(300\text{ K}))/L(300\text{ K})$, is measured here from 12 K to 300 K by a strain gauge method using a commercial strain gauge of length 1 mm. The strain gauge is adhered along the *ab*-plane and along the *c*-axis of the bulk superconductor using gauge cement. The accuracy of the $\Delta L(T)/L$ measurement is better than 10^{-6}.

A.3 Typical results

In this section, typical experimentally-measured thermal conductivity and thermal dilatation for REBaCuO and MgB_2 bulks are given. Most of the experimental data for the bulk (RE)BCO can be downloaded from the following website: http://ikebehp.mat.iwate-u.ac.jp/database.html.

A.3.1 Bulk (RE)BCO

A.3.1.1 Thermal conductivity

The thermal properties of the (RE)BCO bulk depend on the direction of measurement due to its anisotropic structure. The temperature dependence and the absolute value of the thermal conductivity also change depending on the type of RE-ion, the porosity of the material, and the non-superconducting Ag and 211 phases. The thermal conductivity, κ_{ab}, of the *ab*-plane is larger than that along the *c*-axis, κ_c (see section 2.2.1.4).

The thermal conductivity, κ_{ab}, of the *ab*-plane shows a significant upturn below T_c, takes a maximum and then decreases with decreasing *T*. On the other hand, the thermal conductivity, κ_c, along the *c*-direction increases slightly with decreasing *T* without any anomaly around T_c and takes a maximum and then decreases.

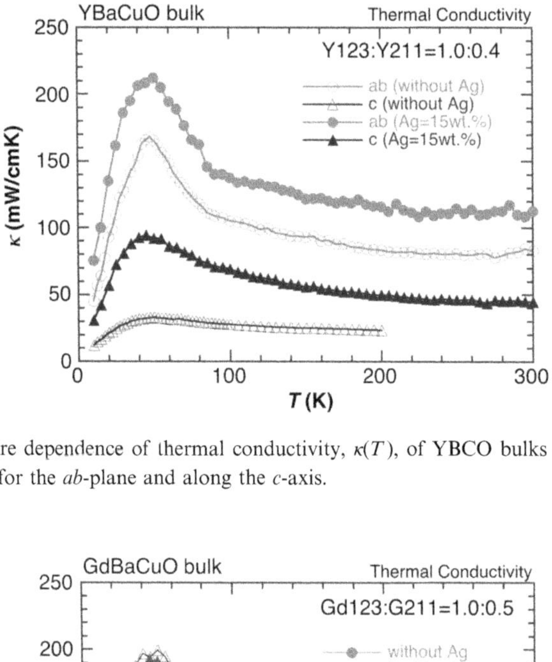

Figure A2. Temperature dependence of thermal conductivity, $\kappa(T)$, of YBCO bulks (Y123:Y211 = 1.0:0.4) with and without Ag, for the *ab*-plane and along the *c*-axis.

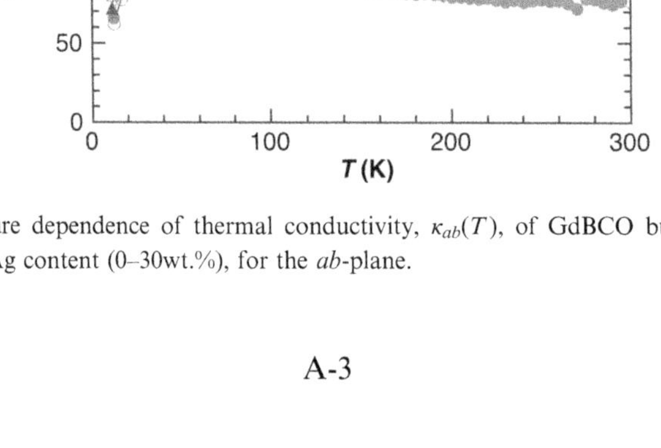

Figure A3. Temperature dependence of thermal conductivity, $\kappa_{ab}(T)$, of GdBCO bulks (Gd123:Gd211 = 1.0:0.5) with various Ag content (0–30wt.%), for the *ab*-plane.

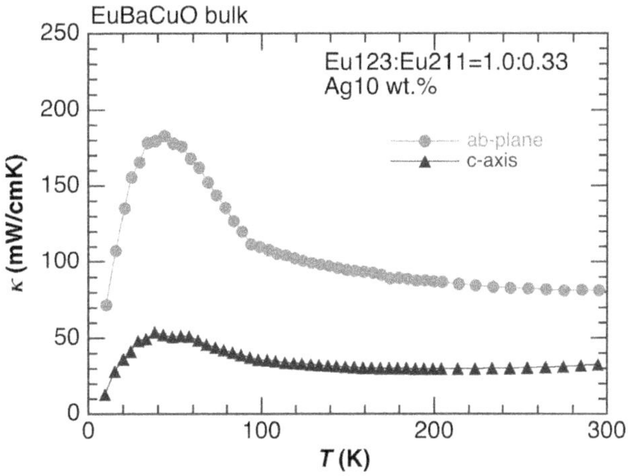

Figure A4. Temperature dependence of thermal conductivity, $\kappa(T)$, of a EuBCO bulk (Eu123:Eu211 = 1.0:0.33) with 10wt.% Ag, for the *ab*-plane and along the *c*-axis.

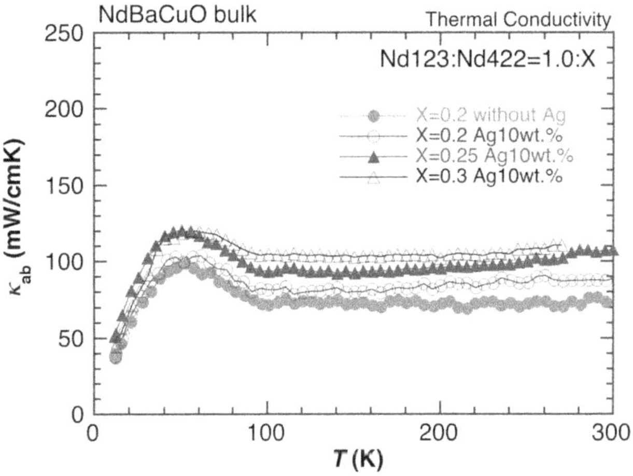

Figure A5. Temperature dependence of thermal conductivity, $\kappa_{ab}(T)$, of NdBCO bulks (Nd123:Nd422 = 1.0: X) with and without 10wt.%Ag, for the *ab*-plane.

The anisotropy ratio, κ_{ab}/κ_c, is around 3 to 5, which has a slight dependence on the RE-ion, porosity, and Ag and 211 phases (figures A2–A7).

The origin of the increased κ_{ab} below T_c is explained by phonon contribution. Assuming that the phonon is the main heat carrier over the entire temperature range, the phonon mean free path becomes long below T_c due to the decrease of phonon scattering by the decrease of normal electrons (or the increase of quasi-particles).

κ_{ab} for YBCO bulks is larger, compared to that of other (RE)BCO bulks, which is closely related to the (RE-Ba) substitution during the crystal growth. The κ_{ab} of DyBCO is comparatively much smaller, which is preferable for current lead applications.

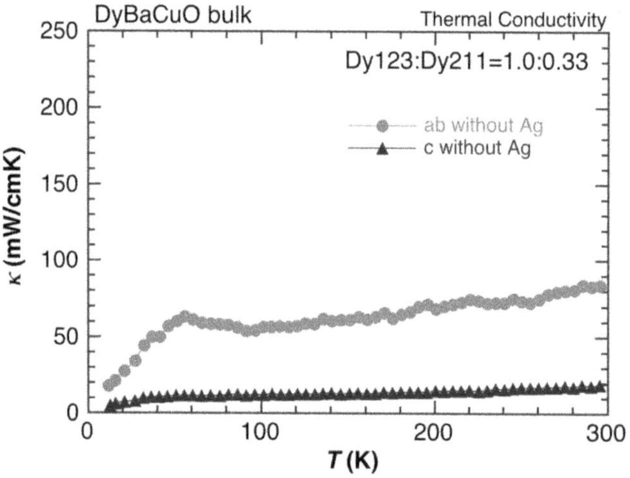

Figure A6. Temperature dependence of thermal conductivity, $\kappa(T)$, of a DyBCO bulk (Dy123:Dy211 = 1.0:0.33) without Ag, for the *ab*-plane and along the *c*-axis.

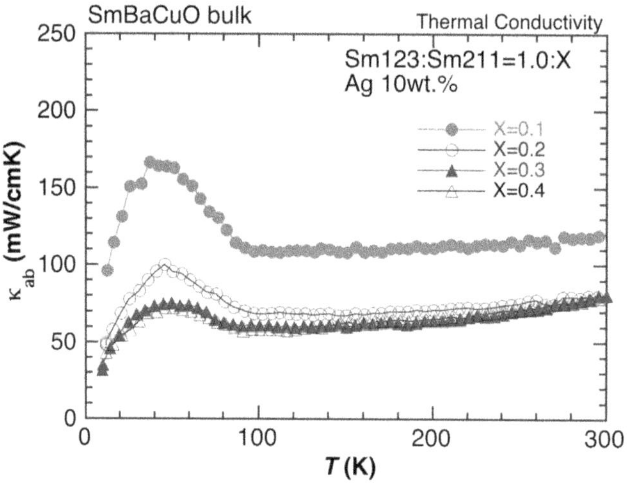

Figure A7. Temperature dependence of thermal conductivity, $\kappa_{ab}(T)$, of SmBCO bulks (Sm123:Sm211 = 1.0:X) with 10wt.% Ag for *ab*-plane.

The absolute value of κ_{ab} decreases with increasing RE-211 phase content and increases with increasing Ag content.

A.3.1.2 Thermal conductivity in magnetic fields

Here, κ_{ab} of a SmBCO bulk (Sm123:Sm211 = 1.0: 0.1) (Ag = 10wt.%) is presented when under the influence of a magnetic field. Figures A8(a) and A8(b) show κ_{ab} for field-cooling runs with the applied field, *B*, parallel ($Q \perp B$, $ab // B$) and perpendicular ($Q // B$, $ab \perp B$) to the *ab*-plane, respectively. In both figures, κ_{ab} above T_c is independent of the applied magnetic field. In figure A8(b), the κ_{ab} enhancement

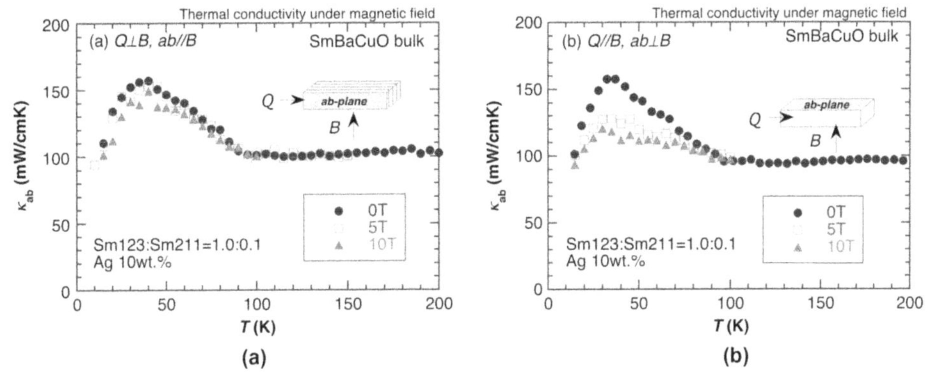

Figure A8. $\kappa_{ab}(T)$ of the SmBCO bulk (Sm123:Sm211 = 1.0:0.1) for field-cooling runs with the applied field B parallel ($Q \perp B$, $ab//$B) and perpendicular ($Q//B$, $ab \perp B$) to the *ab*-plane.

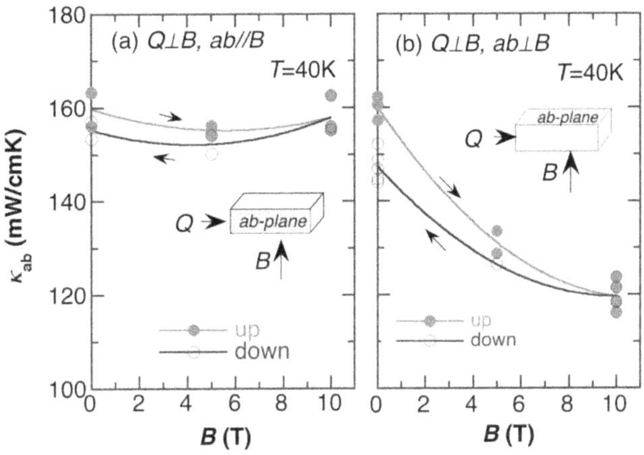

Figure A9. The magnetic field dependence of κ_{ab} of a SmBCO bulk (Sm123:Sm211 = 1.0:0.1) at 40 K, for which the applied field, B, was swept up and down, up to 10 T.

below T_c is drastically suppressed with increasing B perpendicular to the *ab*-plane. This reduction can be understood as caused by enhanced phonon scattering by the quasi-particles in the vortex cores. For the application of B parallel to the *ab*-plane, as shown in figure A8(a), however, the reduction of the κ_{ab} enhancement is barely discernible. In this case, the vortices parallel to the *ab*-plane hardly impede the heat flow. A similar anisotropic effect of the magnetic field on $\kappa_{ab}(T)$ has been observed for YBCO and Bi-2212 single- and poly-crystals.

Figure A9 shows the magnetic field dependence of κ_{ab} at 40 K, where B was swept up and down, up to 10 T. For the $B \perp ab$-plane setting, κ_{ab}(40K) promptly decreases with increasing magnetic field and partly recovers with decreasing field; very clear hysteretic behaviour of κ_{ab} is observed. The κ_{ab}(40K) value at 0 T after applying the field is 8% smaller than the initial one. This suggests that the trapped magnetic field scatters the phonon transport even after removing the field. The trapped magnetic field for the $B \perp ab$-plane setting is estimated to be about 2 T from the data for the increasing virgin run. On the other hand, for the *B*//*ab*-plane setting, the field dependence of κ_{ab}(40K) is quite small up to the maximum applied field of 10 T.

A.3.1.3 Thermal dilatation

The thermal dilatation, d*L*/*L*, of the (RE)BCO bulk changes depending on the measurement direction and the presence of Ag and RE-211 phases (figures A10–A13). The d*L*/*L* value along the *c*-axis is about twice as large as that of the *ab*-plane and slightly increases with increasing Ag content. Roughly speaking, no difference in d*L*/*L* is observed for different RE-ions.

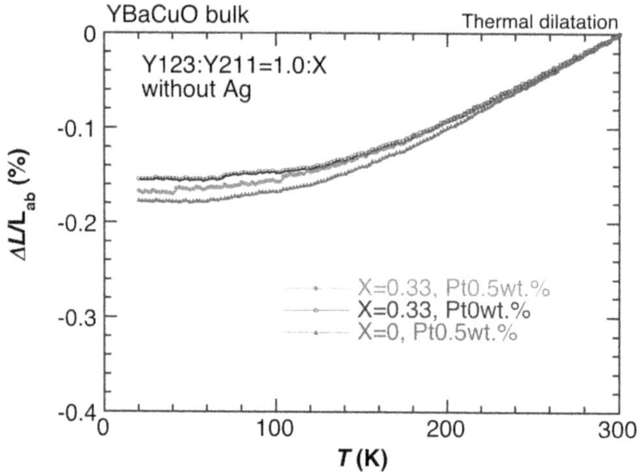

Figure A10. Temperature dependence of thermal dilatation, d*L*/*L*, of YBCO bulks (Y123:Y211 = 1.0:X) without Ag, for the *ab*-plane.

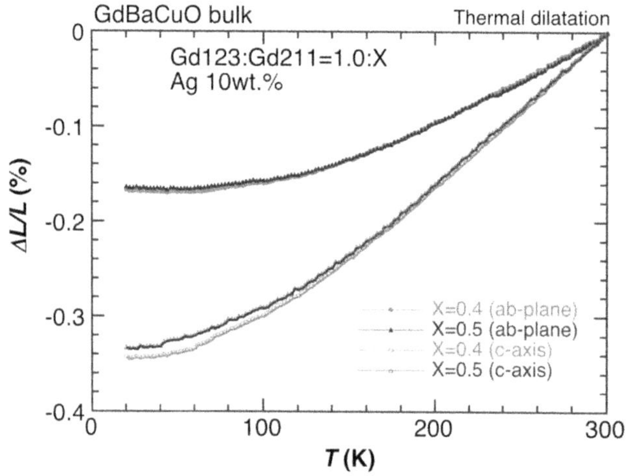

Figure A11. Temperature dependence of thermal dilatation, d*L*/*L*, of GdBCO bulks (Gd123:Gd211 = 1.0:X) with 10wt.% Ag, for the *ab*-plane and along the *c*-axis.

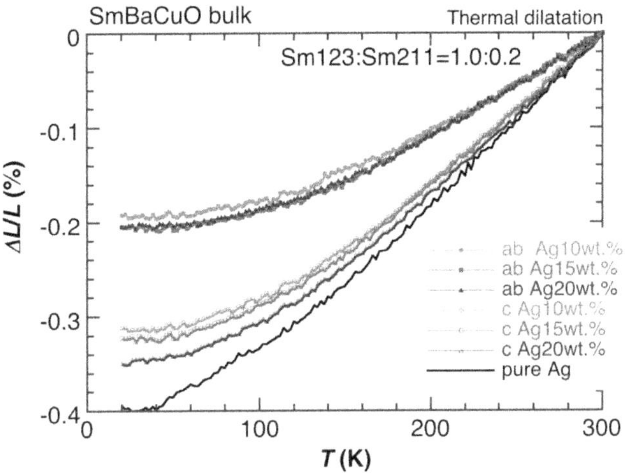

Figure A12. Temperature dependence of thermal dilatation, d*L*/*L*, of SmBCO bulks (Sm123:Sm211 = 1.0:0.2) with Ag from 10 to 20 wt.%, for the *ab*-plane and along the *c*-axis.

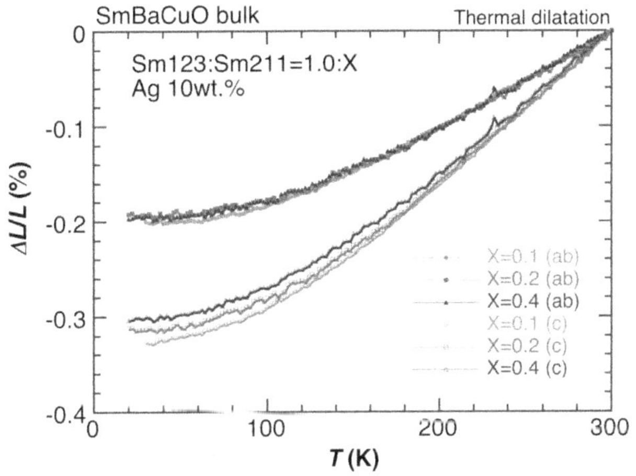

Figure A13. Temperature dependence of thermal dilatation, d*L*/*L*, of SmBCO bulks (Sm123:Sm211 = 1.0:X) with 10wt.% Ag, for the *ab*-plane and along the *c*-axis.

A.3.2 Bulk MgB_2

MgB_2 bulk is a polycrystalline material, which can be fabricated by several different techniques as described in section 1.3.2. The temperature dependence and the absolute value of the thermal conductivity of MgB_2 bulks change depending on the relative density of the material and additive elements (figure A14). The thermal dilatation of MgB_2 is smaller than that for the *ab*-plane of (RE)BCO bulks (figure A15).

A.3.2.1 Thermal conductivity

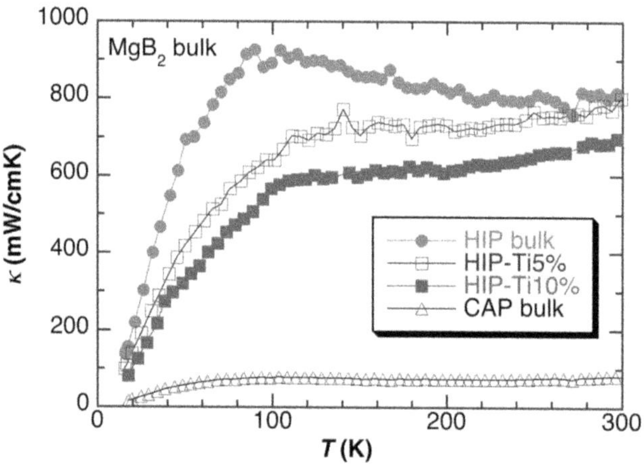

Figure A14. Temperature dependence of thermal conductivity, $\kappa(T)$, of MgB_2 bulks fabricated by different methods: hot isostatic pressing (HIP) with varying Ti content and a capsule method (CAP).

A.3.2.2 Thermal dilatation

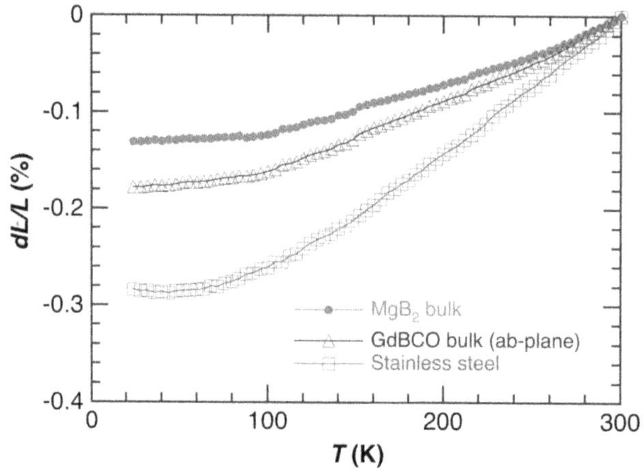

Figure A15. Temperature dependence of thermal dilatation, dL/L, of MgB_2 bulks fabricated by the HIP method, compared with results for a GdBCO bulk (*ab*-plane) and stainless steel.

Further reading

(1) Thermal properties of bulk (RE)BCO

[1] Ikebe M, Fujishiro H, Naito T, Noto K, Kohayashi S and Yoshizawa S 1994 Thermal conductivity of YBCO(123) and YBCO(211) mixed crystals prepared by MMTG *Cryogenics* **34** 57–61

[2] Fujishiro H, Ikebe M, Naito T, Noto K, Kohayashi S and Yoshizawa S 1994 Anisotropic thermal diffusivity and conductivity of YBCO(123) and YBCO(211) mixed crystals I *Jpn. J. Appl. Phys.* **33** 4965–70

[3] Ikebe M, Fujishiro H, Naito T, Matsukawa M and Noto K 1994 Anisotropic thermal diffusivity and conductivity of YBCO(123) and YBCO(211) mixed crystals II *Jpn. J. Appl. Phys.* **33** 6157–9

[4] Fujishiro H, Ikebe M, Kohayashi S, Noto K and Yokoyama K 2003 Additional effect of Ag and 211 phase on thermal expansion of REBaCuO bulk superconductor (RE = Sm, Y) *Cryogenics* **43** 477–81

[5] Fujishiro H, Teshima H, Ikebe M and Noto K 2003 Thermal conductivity of YBaCuO bulk superconductors under applied field : effect of content and size of Y211 phase *Physica* C 392–396 P1 171–4

[6] Noto K, Oka T, Yokoyama K, Katagiri K, Fujishiro H, Okada H, Nakazawa H, Muralidhar M and Murakami M 2003 Thermal and mechanical properties of high T_c bulk superconductors and their applications Physica C **392–396** 677–83

[7] Fujishiro H, Katagiri K, Murakami A, Yoshino Y and Noto K 2005 Database for thermal and mechanical properties of REBaCuO bulks *Physica* C **426–431** 699–704

[8] Fujishiro H, Nariki S and Murakami M 2006 Thermal conductivity and thermoelectric power of DyBaCuO dulk superconductor *Supercond. Sci. Technol.* **19** S447–50

[9] Fujishiro H, Ikebe M, Teshima H and Hirano H 2006 Low-thermal-conductive DyBaCuO bulk superconductor for current lead application *IEEE Trans. Appl. Supercond.* **16** 1080–3

[10] Naito T, Fujishiro H, Iida K, Nariki S and Murakami M 2008 Thermal conductivity of Er-Ba-Cu-O and Ho-Ba-Cu-O superconducting bulks *Supercond. Sci. Technol.* **21** 085001

[11] Naito T, Fujishiro H, Iida K and Murakami M 2008 Anisotropic thermal conductivity of Er-Ba-Cu-O bulk superconductors *Physica* C **468** 1428–30

(2) Thermal properties of bulk MgB_2

[12] Naito T, Mochizuki H, Fujishiro H and Teshima H 2016 Trapped magnetic-field properties of prototype for Gd-Ba-Cu-O/MgB_2 hybrid-type superconducting bulk magnet *Supercond. Sci. Technol.* **29** 034005

[13] Cavallin T, Young E A, Beduz C, Yang Y and Giunchi G 2007 Thermal conductivity of bulk MgB_2 produced by infiltration of different boron powders *IEEE Trans. Appl. Supercond.* **17** 2770–3

[14] Zou J, Ainslie M D, Fujishiro H, Bhagurkar A G, Naito T, Hari Babu N, Fagnard J-F, Vanderbemden P and Yamamoto A 2015 Numerical modeling and comparison of MgB_2 bulks fabricated by HIP and infiltration growth magnetized by field cooling *Supercond. Sci. Technol.* **28** 075009

[15] Fujishiro H, Naito T and Yoshida T 2014 Numerical simulation of trapped field on MgB_2 bulk disk magnetized by field cooling *Supercond. Sci. Technol.* **27** 065019

(3) Techniques for measuring thermal properties

[16] Fujishiro H, Okamoto T and Hirose K 2001 Thermal contact resistance between high-T_c superconductor and Cu *Physica* C **357–360** 785–8

[17] Ikebe M, Fujishiro H, Naito T and Noto K 1994 Simultaneous measurement of thermal diffusivity and conductivity applied to Bi-2223 ceramic superconductors *J. Phys. Soc. Jpn.* **63** 3107–14

[18] Fujishiro H, Ikebe M, Nakasato K and Naito T 1996 Proposal of three terminal method for low temperature thermal diffusivity measurement *Czech. J. Phys.* **46** Suppl. S5 2761–2

[19] Fujishiro H, Okamoto T, Ikebe M and Hirose K 2001 A new method for simultaneous determination of anisotropic thermal conductivities based on two-dimensional analyses *Jpn. J. Appl. Phys.* **40** 388–92

[20] Fujishiro H, Okamoto T and Hirose K 2001 Thermal contact resistance between high-T_c superconductor and copper *Physica* C **357–360** 785–8

Lightning Source UK Ltd.
Milton Keynes UK
UKHW032033191219
355695UK00006B/300/P

9 780750 319577